Statistical Learning with Math and Python

Joe Suzuki

Statistical Learning with Math and Python

100 Exercises for Building Logic

 Springer

Joe Suzuki
Graduate School of Eng Sci
Osaka University
Toyonaka, Osaka, Japan

ISBN 978-981-15-7876-2   ISBN 978-981-15-7877-9   (eBook)
https://doi.org/10.1007/978-981-15-7877-9

This Springer imprint is published by the registered company Springer Nature Singapore Pte Ltd.
The registered company address is: 152 Beach Road, #21-01/04 Gateway East, Singapore 189721,
Singapore

Preface

I am currently with the Statistics Laboratory at Osaka University, Japan. I often meet with data scientists who are engaged in machine learning and statistical analyses for research collaborations and introducing my students to them. I recently found out that almost all of them think that (mathematical) logic rather than knowledge and experience is the most crucial ability for grasping the essence in their jobs. Our necessary knowledge is changing every day and can be obtained when needed. However, logic allows us to examine whether each item on the Internet is correct and follow any changes; without it, we might miss even chances.

In 2016, I started teaching statistical machine learning to the undergraduate students of the Mathematics Department. In the beginning, I was mainly teaching them what (statistical) machine learning (ML) is and how to use it. I explained the procedures of ML, such as logistic regression, support vector machines, k-means clustering, etc., by showing figures and providing intuitive explanations. At the same time, the students tried to understand ML by guessing the details. I also showed the students how to execute the ready-made functions in several R packages without showing the procedural details; at the same time, they understood how to use the R packages as black boxes.

However, as time went by, I felt that this manner of teaching should be changed. In other non-ML classes, I focus on making the students consider extending the ideas. I realized that they needed to understand the essence of the subject by mathematically considering problems and building programs. I am both a mathematician and an R/Python programmer and notice the importance of instilling logic inside each student. The basic idea is that the students see that both theory and practice meet and that using logic is necessary.

I was motivated to write this book because I could not find any other book that was inspired by the idea of "instilling logic" in the field of ML.

The closest comparison is "Introduction to Statistical Learning: with Application in R" (ISLR) by Gareth James, Daniela Witten, Trevor Hastie, and Robert Tibshirani (Springer), which is the most popular book in this field. I like this book and have used it for the aforementioned class. In particular, the presentation in the book is splendid (abundant figures and intuitive explanations). I followed this style when writing this book. However, ISLR is intended for a beginner audience. Compared with ISLR, this book (SLMP) focuses more on mathematics and programming, although the contents are similar: linear regression, classification, information criteria, regularizations, decision trees, support vector machine, and unsupervised learning.

Another similar book is "The Elements of Statistical Learning" (ESL) by Trevor Hastie, Robert Tibshirani, and Jerome Friedman (Springer), which provides the most reliable knowledge on statistical learning. I often use it when preparing for my classes. However, the volume of information in ESL is large, and it takes at least 500–1000 h to read it through, although I do recommend reading the book. My book, SLMP, on the other hand, takes at most 100 h, depending on the reader's baseline ability, and it does not assume the reader has any knowledge of ML. After reading SLMP, it takes at most 300–500 h to read through ESL because the reader will have enough logic to easily understand ESL. ESL contains many equations and procedures but no programming codes.

In this sense, SLMP focuses on both mathematics and programming more than ISLR and ESL. I sincerely wish that the reader of SLMP will develop both logic and statistical learning knowledge.

What Makes SLMP Unique?

I have summarized the features of this book as follows.

1. Developing logic
   To grasp the essence of the subject, we mathematically formulate and solve each ML problem and build those programs. The SLMP instills "logic" in the minds of the readers. The reader will acquire both the knowledge and ideas of ML, so that even if new technology emerges, they will be able to follow the changes smoothly. After solving the 100 problems, most of the students would say "I learned a lot."
2. Not just a story
   If programming codes are available, you can immediately take action. It is unfortunate when an ML book does not offer the source codes. Even if a package is available, if we cannot see the inner workings of the programs, all we can do is input data into those programs. In SLMP, the program codes are available for most of the procedures. In cases where the reader does not understand the math, the codes will help them understand what it means.
3. Not just a how-to book: an academic book written by a university professor
   This book explains how to use the package and provides examples of executions for those who are not familiar with them. Still, because only the inputs and outputs are visible, we can only see the procedure as a black box. In this sense, the reader will have limited satisfaction because they will not be able to obtain

the essence of the subject. SLMP intends to show the reader the heart of ML and is more of a full-fledged academic book.

4. Solve 100 exercises: problems are improved with feedback from university students

The exercises in this book have been used in university lectures and have been refined based on feedback from students. The best 100 problems were selected. Each chapter (except the exercises) explains the solutions, and you can solve all of the exercises by reading the book.

5. Self-contained

All of us have been discouraged by phrases such as "for the details, please refer to the literature XX." Unless you are an enthusiastic reader or researcher, nobody will seek out those references. In this book, we have presented the material in such a way that consulting external references is not required. Additionally, the proofs are simple derivations, and the complicated proofs are given in the appendices at the end of each chapter. SLMP completes all discussions, including the appendices.

6. Readers' pages: questions, discussion, and program files

The reader can ask any question on the book's Facebook page (https://bayesnet. org/books). Additionally, all programs and data can be downloaded from http:// bitbucket.org/prof-joe (thus, you do not have to copy the programs from the book).

7. Linear algebra

One of the bottlenecks in learning ML and statistics is linear algebra. Except for books for researchers, a few books assume the reader has knowledge of linear algebra, and most books cannot go into the details of this subject. Therefore, SLMP contains a summary of linear algebra. This summary is only 17 pages and is not just an example, but it provides all the proofs. If you already know linear algebra, then you can skip it. However, if you are not confident in the subject, you can read in only one day.

How to Use This Book

Each chapter consists of problems, their explanation (body), and an appendix (proof, program). You can start reading the body and solve the problem. Alternatively, you might want to solve the 100 exercises first and consult the body if necessary. Please read through the entire book until the end.

When used in a lecture, I recommend that the teacher organizes the class into 12, 90 min lectures (or a 1000 min course) as follows: 3 lectures for Chap. 1, 2 lectures for Chap. 6, and 1 lecture for each of the other chapters. You may ask the students to complete the 100 exercises. If you read the text carefully, you will be able to answer any of their questions. I think that the entire book can be fully read in about 12 lectures total.

Acknowledgments

The author wishes to thank Yuske Inaoka, Tianle Yang, Ryosuke Shinmura, Junichi Maruyama, and Kazuya Morishita for checking the manuscript. This English book is largely based on the Japanese book published by Kyoritsu Shuppan Co., Ltd, in 2020. The author would like to thank Kyoritsu Shuppan Co., Ltd, for their generosity. The author also appreciates Ms. Mio Sugino, Springer, for preparing the publication and providing advice on the manuscript.

Osaka, Japan

Joe Suzuki

May 2021

Contents

Chapter 1
Linear Algebra

Abstract Linear algebra is the basis of logic constructions in any science. In this chapter, we learn about inverse matrices, determinants, linear independence, vector spaces and their dimensions, eigenvalues and eigenvectors, orthonormal bases and orthogonal matrices, and diagonalizing symmetric matrices. In this book, to understand the essence concisely, we define ranks and determinants based on the notion of Gaussian elimination and consider linear spaces and their inner products within the range of the Euclidean space and the standard inner product. By reading this chapter, the readers should solve the reasons why.

1.1 Inverse Matrix

First, we consider solving the problem $Ax = b$ w.r.t. $x \in \mathbb{R}^n$ for $A \in \mathbb{R}^{m \times n}$, $b \in \mathbb{R}^m$. We refer to $A \in \mathbb{R}^{m \times n}$ and $[A|b] \in \mathbb{R}^{m \times (n+1)}$ as a coefficient matrix and an extended coefficient matrix, respectively. We write $A \sim B$ when A can be transformed into $B \in \mathbb{R}^{m \times n}$ via the three elementary row operations below:

Operation 1 divides one whole row by a nonzero constant.
Operation 2 exchanges two rows.
Operation 3 adds one row multiplied by a constant to another row.

Example 1 $\begin{cases} 2x + 3y = 8 \\ x + 2y = 5 \end{cases} \iff \begin{cases} x = 1 \\ y = 2 \end{cases}$ is equivalent to $\begin{bmatrix} 2 & 3 & | & 8 \\ 1 & 2 & | & 5 \end{bmatrix} \sim$ $\begin{bmatrix} 1 & 0 & | & 1 \\ 0 & 1 & | & 2 \end{bmatrix}$.

Example 2 $\begin{cases} 2x - y + 5z = -1 \\ y + z = 3 \\ x + 3z = 1 \end{cases} \iff \begin{cases} x + 3z = 1 \\ y + z = 3 \end{cases}$ is equivalent to $\begin{bmatrix} 2 & -1 & 5 & | & -1 \\ 0 & 1 & 1 & | & 3 \\ 1 & 0 & 3 & | & 1 \end{bmatrix} \sim \begin{bmatrix} 1 & 0 & 3 & | & 1 \\ 0 & 1 & 1 & | & 3 \\ 0 & 0 & 0 & | & 0 \end{bmatrix}$.

© The Author(s), under exclusive license to Springer Nature Singapore Pte Ltd. 2021
J. Suzuki, *Statistical Learning with Math and Python*,
https://doi.org/10.1007/978-981-15-7877-9_1

Example 3 $\begin{cases} 2x - y + 5z = 0 \\ y + z = 0 \\ x + 3z = 0 \end{cases}$ $\iff$ $\begin{cases} x + 3z = 0 \\ y + z = 0 \end{cases}$ is equivalent to $\begin{bmatrix} 2 & -1 & 5 \\ 0 & 1 & 1 \\ 1 & 0 & 3 \end{bmatrix}$

$\sim \begin{bmatrix} 1 & 0 & 3 \\ 0 & 1 & 1 \\ 0 & 0 & 0 \end{bmatrix}$.

We refer to the first nonzero element in each of the nonzero rows as the main element of the row and the matrix satisfying the following conditions as the canonical form:

- All-zero rows are positioned in the lowest rows.
- The main element is one unless the row is all zeros.
- The lower the row, the closer to the right the position of the main element is.
- For each column, all but the main elements are zero.

Example 4 The following matrices are in the canonical form in which ① is the main element:

$$\begin{bmatrix} 0 & ① & 3 & 0 & 2 \\ 0 & 0 & 0 & ① & 1 \\ 0 & 0 & 0 & 0 & 0 \end{bmatrix}, \begin{bmatrix} ① & 0 & 1 & 4 & 0 & -1 \\ 0 & ① & 7 & -4 & 0 & 1 \\ 0 & 0 & 0 & 0 & ① & 3 \end{bmatrix}, \begin{bmatrix} 0 & ① & 0 & 0 & 2 & 3 \\ 0 & 0 & 0 & 0 & 0 & 0 \\ 0 & 0 & 0 & 0 & 0 & 0 \end{bmatrix},$$

$$\begin{bmatrix} 0 & 0 & ① & 0 & 2 & 0 \\ 0 & 0 & 0 & 0 & 0 & ① \\ 0 & 0 & 0 & 0 & 0 & 0 \end{bmatrix}.$$

For an arbitrary $A \in \mathbb{R}^{m \times n}$, the canonical form is unique, which will be proven at the end of Sect. 1.3. We refer to any procedure computing the canonical form based on the three above operations as Gaussian elimination. In particular, we refer to the number of main elements in matrix A as the rank of A (i.e., rank(A)). From the definition, the rank of A does not exceed the minimum of m, n.

If matrix $A \in \mathbb{R}^{n \times n}$ is square and its canonical form is the unit matrix $I \in \mathbb{R}^{n \times n}$, then we say that A is nonsingular. For square matrices $A, B \in \mathbb{R}^{n \times n}$, if $[A|I] \sim [I|B]$, then the extended coefficient matrix $[I|B]$ is the canonical form of $[A|I]$. In such a case, we say that either A or B is the inverse matrix of the other and write $A^{-1} = B$ and $B^{-1} = A$. The relation $[A|I] \sim [I|B]$ implies that $AX = I \iff B = X$. In fact, if we write $X = [x_1, \ldots, x_n], B = [b_1, \ldots, b_n] \in \mathbb{R}^{n \times n}$, then the relation implies that $Ax_i = e_i \iff b_i = x_i$ for $i = 1, \ldots, n$, where $e_i \in \mathbb{R}^n$ is the unit vector in which the i-th element is one and the other elements are zero.

Example 5 For matrix $A = \begin{bmatrix} 1 & 2 & 1 \\ 2 & 3 & 1 \\ 1 & 2 & 2 \end{bmatrix}$, we have

$$\left[\begin{array}{ccc|ccc} 1 & 2 & 1 & 1 & 0 & 0 \\ 2 & 3 & 1 & 0 & 1 & 0 \\ 1 & 2 & 2 & 0 & 0 & 1 \end{array}\right] \sim \left[\begin{array}{ccc|ccc} 1 & 0 & 0 & -4 & 2 & 1 \\ 0 & 1 & 0 & 3 & -1 & -1 \\ 0 & 0 & 1 & -1 & 0 & 1 \end{array}\right].$$

If we look at the left half on both sides, we can see that $\begin{bmatrix} 1 & 2 & 1 \\ 2 & 3 & 1 \\ 1 & 2 & 2 \end{bmatrix} \sim$

$\begin{bmatrix} 1 & 0 & 0 \\ 0 & 1 & 0 \\ 0 & 0 & 1 \end{bmatrix}$, which means that A is nonsingular. Therefore, we can write

$$\begin{bmatrix} 1 & 2 & 1 \\ 2 & 3 & 1 \\ 1 & 2 & 2 \end{bmatrix}^{-1} = \begin{bmatrix} -4 & 2 & 1 \\ 3 & -1 & -1 \\ -1 & 0 & 1 \end{bmatrix}$$

$$\begin{bmatrix} -4 & 2 & 1 \\ 3 & -1 & -1 \\ -1 & 0 & 1 \end{bmatrix}^{-1} = \begin{bmatrix} 1 & 2 & 1 \\ 2 & 3 & 1 \\ 1 & 2 & 2 \end{bmatrix},$$

and we have the following relation:

$$\begin{bmatrix} 1 & 2 & 1 \\ 2 & 3 & 1 \\ 1 & 2 & 2 \end{bmatrix}\begin{bmatrix} -4 & 2 & 1 \\ 3 & -1 & -1 \\ -1 & 0 & 1 \end{bmatrix} = \begin{bmatrix} 1 & 0 & 0 \\ 0 & 1 & 0 \\ 0 & 0 & 1 \end{bmatrix}.$$

On the other hand, although the solution $x = x^*$ of $Ax = b$ with $x \in \mathbb{R}^n$ and $b \in \mathbb{R}^n$ can be obtained by $A^{-1}b$, we may obtain via $[A|b] \sim [I|x^*]$, which means we find $x = x^*$, without computing A^{-1}.

1.2 Determinant

We define the determinant $\det(A)$ for a square matrix A as follows. If the canonical form is not the unit matrix, which means that A is singular, we set $\det(A) := 0$. If A is the unit matrix I, then we set $\det(A) = 1$. Suppose that A is nonsingular and is not the unit matrix.

If we repeatedly apply the third elementary row operation, we obtain a matrix such that each row and each column contain exactly one nonzero element. Then, if we repeat the second operation, we obtain a diagonal matrix. Finally, if we repeat the third operation, then we obtain the unit matrix.

When computing the determinant $\det(A)$, we execute the reverse procedure from the unit matrix $A = I$ to the original A:

Step 1   Multiply $\det(A)$ by α_i if the i-th row is multiplied by α_i.
Step 2   Multiply $\det(A)$ by -1 if the $i \neq j$ rows are exchanged.

Step 3   Do not change det(A) if the j-th row multiplied by β_j is subtracted from the i-th row.

We define det(A) as the final obtained value after executing the three steps. Let m be how many times we implement Step 2 (multiplying by -1). Then, we have det(A) = $(-1)^m \prod_{i=1}^n \alpha_i$.

Proposition 1 *For a matrix $A \in \mathbb{R}^{n \times n}$, A is nonsingular $\Longleftrightarrow$ rank(A) = n $\Longleftrightarrow$ det(A) $\neq$ 0.*

Example 6 If an all-zero row appears, which means that the determinant is zero, we may terminate the procedure. For the matrix below, the determinant is six because we exchange the rows once at the beginning.

$$\begin{bmatrix} 0 & 1 & 1 \\ 0 & -1 & 5 \\ 1 & 0 & 4 \end{bmatrix} \sim \begin{bmatrix} 1 & 0 & 4 \\ 0 & -1 & 5 \\ 0 & 1 & 1 \end{bmatrix} \sim \begin{bmatrix} 1 & 0 & 4 \\ 0 & -1 & 5 \\ 0 & 0 & 6 \end{bmatrix} \sim \begin{bmatrix} 1 & 0 & 0 \\ 0 & -1 & 0 \\ 0 & 0 & 6 \end{bmatrix}.$$

In the following example, since an all-zero row appears during Gaussian elimination, the determinant is zero.

$$\begin{bmatrix} 2 & -1 & 5 \\ 0 & 1 & 1 \\ 1 & 0 & 3 \end{bmatrix} \sim \begin{bmatrix} 2 & -1 & 5 \\ 0 & 1 & 1 \\ 0 & 1/2 & 1/2 \end{bmatrix} \sim \begin{bmatrix} 2 & 0 & 6 \\ 0 & 1 & 1 \\ 0 & 0 & 0 \end{bmatrix}.$$

In general, for a 2×2 matrix, if $a \neq 0$, we have

$$\begin{vmatrix} a & b \\ c & d \end{vmatrix} = \begin{vmatrix} a & b \\ 0 & d - \dfrac{bc}{a} \end{vmatrix} = ad - bc,$$

and even if $a = 0$, the determinant is $ad - bc$.

$$\begin{vmatrix} 0 & b \\ c & d \end{vmatrix} = -\begin{vmatrix} c & d \\ 0 & b \end{vmatrix} = -bc.$$

Therefore, $ad - bc \neq 0$ is the condition for A to be nonsingular, and from

$$\begin{bmatrix} a & b \\ c & d \end{bmatrix} \cdot \frac{1}{ad - bc} \begin{bmatrix} d & -b \\ -c & a \end{bmatrix} = \begin{bmatrix} 1 & 0 \\ 0 & 1 \end{bmatrix},$$

we have

$$\begin{bmatrix} a & b \\ c & d \end{bmatrix}^{-1} = \frac{1}{ad - bc} \begin{bmatrix} d & -b \\ -c & a \end{bmatrix}.$$

On the other hand, for 3×3 matrices, if $a \neq 0$ and $ae \neq bd$, we have

$$
\begin{vmatrix} a & b & c \\ d & e & f \\ g & h & i \end{vmatrix} = \begin{vmatrix} a & b & c \\ 0 & e - \dfrac{bd}{a} & f - \dfrac{cd}{a} \\ 0 & h - \dfrac{bg}{a} & i - \dfrac{cg}{a} \end{vmatrix}
$$

$$
= \begin{vmatrix} a & 0 & c - \dfrac{b}{e - bd/a} \cdot \left(f - \dfrac{cd}{a} \right) \\ 0 & e - \dfrac{bd}{a} & f - \dfrac{cd}{a} \\ 0 & 0 & i - \dfrac{cg}{a} - \dfrac{h - bg/a}{e - bd/a} \left(f - \dfrac{cd}{a} \right) \end{vmatrix}
$$

$$
= \begin{vmatrix} a & 0 & 0 \\ 0 & \dfrac{ae - bd}{a} & 0 \\ 0 & 0 & \dfrac{aei + bfg + cdh - ceg - bdi - afh}{ae - bd} \end{vmatrix}
$$

$$
= aei + bfg + cdh - ceg - bdi - afh .
$$

Even if either $a = 0$ or $ae = bd$ holds, we can see that the determinant is $aei + bfg + cdh - ceg - bdi - afh$.

Proposition 2 *For square matrices A and B of the same size, we have $\det(AB) = \det(A)\det(B)$ and $\det(A^T) = \det(A)$.*

For the proof, see the Appendix at the end of this chapter. The equation $\det(A^T) = \det(A)$ in Proposition 2 means that we may apply the following rules to obtain the determinant $\det(A)$:

Step 2′   Multiply $\det(A)$ by -1 if the $i \neq j$ columns are exchanged.

Step 3′   Do not change $\det(A)$ if the j-th column multiplied by β_j is subtracted from the i-th columns.

Example 7 (Vandermonde's Determinant)

$$
\begin{vmatrix} 1 & a_1 & \cdots & a_1^{n-1} \\ \vdots & \vdots & \ddots & \vdots \\ 1 & a_n & \cdots & a_n^{n-1} \end{vmatrix} = (-1)^{n(n-1)/2} \prod_{1 \le i < j \le n} (a_i - a_j). \tag{1.1}
$$

In fact, if $n = 1$, both sides are one, and the claim holds. If we assume the claim for $n = k - 1$, then for $n = k$, the left-hand side of (1.1) is

$$
\begin{vmatrix}
1 & a_1 & \cdots & a_1^{k-2} & a_1^{k-1} \\
0 & a_2 - a_1 & \cdots & a_k^{k-2} - a_1^{k-2} & a_k^{k-1} - a_1^{k-1} \\
\vdots & \vdots & \ddots & \vdots & \vdots \\
0 & a_k - a_1 & \cdots & a_k^{k-2} - a_1^{k-2} & a_k^{k-1} - a_1^{k-1}
\end{vmatrix}
$$

$$
=
\begin{vmatrix}
1 & 0 & \cdots & 0 & 0 \\
0 & a_2 - a_1 & \cdots & (a_2 - a_1)a_2^{k-2} & (a_2 - a_1)a_2^{k-1} \\
\vdots & \vdots & \ddots & \vdots & \vdots \\
0 & a_k - a_1 & \cdots & (a_k - a_1)a_k^{k-2} & (a_k - a_1)a_k^{k-1}
\end{vmatrix}
$$

$$
=
\begin{vmatrix}
a_2 - a_1 & \cdots & (a_2 - a_1)a_2^{k-2} \\
\vdots & \ddots & \vdots \\
a_k - a_1 & \cdots & (a_k - a_1)a_k^{k-2}
\end{vmatrix}
= (a_2 - a_1) \ldots (a_k - a_1)
\begin{vmatrix}
1 & a_2 & \cdots & a_2^{k-2} \\
\vdots & \vdots & \ddots & \vdots \\
1 & a_k & \cdots & a_k^{k-2}
\end{vmatrix}
$$

$$
= (-1)^{k-1}(a_1 - a_2) \ldots (a_1 - a_k) \cdot (-1)^{(k-1)(k-2)/2} \prod_{2 \leq i < j \leq k} (a_i - a_j),
$$

where the left-hand side is obtained by subtracting the first row from the other rows. The first equation is obtained by subtracting the $(j - 1)$-th column multiplied by a_1 from the j-th column for $j = k, k - 1, \ldots, 2$. The third equation is obtained by dividing the rows by constants and multiplying the determinant by the same constants. The last transformation is due to the assumption of induction, and this value coincides with the right-hand side of (1.1). Thus, from induction, we have (1.1).

1.3 Linear Independence

For a matrix $A \in \mathbb{R}^{m \times n}$ with column vectors $a_1, \ldots, a_n \in \mathbb{R}^m$, if the solution of $Ax = 0$ is only $x = 0 \in \mathbb{R}^n$, we say that $a_1, \ldots, a_n$ are linearly independent; otherwise, we say that they are linearly dependent.

Given a set of vectors, we refer to any instance of linear independence or dependence as a linear relation. If $A \sim B$, we can see that the linear relations among the column vectors in A and B are equivalent.

Example 8 For $A = [a_1, a_2, a_3, a_4, a_5]$ and $B = [b_1, b_2, b_3, b_4, b_5]$, $A \sim B$ means $Ax = 0 \iff Bx = 0$.

$$
a_1 = \begin{bmatrix} 1 \\ 1 \\ 3 \\ 0 \end{bmatrix}, a_2 = \begin{bmatrix} 1 \\ 2 \\ 0 \\ -1 \end{bmatrix}, a_3 = \begin{bmatrix} 1 \\ 3 \\ -3 \\ 2 \end{bmatrix}, a_4 = \begin{bmatrix} -2 \\ -4 \\ 1 \\ -1 \end{bmatrix}, a_5 = \begin{bmatrix} -1 \\ -4 \\ 7 \\ 0 \end{bmatrix},
$$

$$
\begin{bmatrix} 1 & 1 & 1 & -2 & -1 \\ 1 & 2 & 3 & -4 & -4 \\ 3 & 0 & -3 & 1 & 7 \\ 0 & -1 & -2 & -1 & 0 \end{bmatrix} \sim \begin{bmatrix} 1 & 0 & -1 & 0 & 2 \\ 0 & 1 & 2 & 0 & -1 \\ 0 & 0 & 0 & 1 & 1 \\ 0 & 0 & 0 & 0 & 0 \end{bmatrix},
$$

$\left. \begin{array}{l} a_1, a_2, a_4 \text{ are linearly independent} \\ a_3 = -a_1 + 2a_2 \\ a_5 = 2a_1 - a_2 + a_4 \end{array} \right\} \iff \left\{ \begin{array}{l} b_1, b_2, b_4 \text{ are linearly independent} \\ b_3 = -b_1 + 2b_2 \\ b_5 = 2b_1 - b_2 + b_4. \end{array} \right.$

We interpret the rank as the maximum number of linearly independent columns in the matrix.

If $a_1, \ldots, a_n \in \mathbb{R}^m$ are linearly independent, none of them can be expressed by any linear combination of the others. If we can express them as $a_i = \sum_{j \neq i} x_j a_j$ for some i, we would have $Ax = \sum_{i=1}^n x_i a_i = 0$, which means that there exists $x \in \mathbb{R}^n$ such that $x_i \neq 0$. On the other hand, if they are linearly dependent, such an $x_i \neq 0$ that $Ax = 0$ exists, and we write $a_i = \sum_{j \neq i} (-x_j / x_i) a_j$. Moreover, even if we define a vector a_{r+1} by a linear combination $a_1, \ldots, a_r$, then $a_1, \ldots, a_r, a_{r+1}$ are linearly dependent. Thus, if we right-multiply $A \in \mathbb{R}^{m \times n}$ by a matrix $B \in \mathbb{R}^{n \times l}$, where B is on the right, then we obtain a matrix AB whose column vectors are $\sum_{i=1}^n a_i b_{i,1}, \ldots, \sum_{i=1}^n a_i b_{i,l}$, which means that the rank (the number of linearly independent vectors) does not exceed the rank of A, i.e., $\text{rank}(AB) \leq \text{rank}(A)$.

When a matrix B is obtained from elementary row operations applied to matrix A, the number of linearly independent row vectors in B does not exceed that in A. Similarly, because A can be obtained from B via elementary row operations, the numbers of linearly independent row vectors are equal, which holds even when B is the canonical form of A. On the other hand, all nonzero rows in the canonical form are linearly independent, and the number of such vectors is the same as that of the main elements. Therefore, the rank is the number of linearly independent row vectors in A as well. Thus, A and its transpose A^T share the same rank. Moreover, the matrix BA obtained by multiplying $B \in \mathbb{R}^{l \times m}$ from right by A has the same rank as $(BA)^T = A^T B^T$, which means that $\text{rank}(BA)$ does not exceed $\text{rank}(A^T)$, which is equal to $\text{rank}(A)$. We summarize the above discussion as follows.

Proposition 3 *For $A \in \mathbb{R}^{m \times n}$ and $B \in \mathbb{R}^{n \times l}$, we have*

$$rank(AB) \leq \min\{rank(A), rank(B)\}$$
$$rank(A^T) = rank(A) \leq \min\{m, n\}.$$

Example 9 From $A = \begin{bmatrix} 2 & 3 \\ 1 & 2 \end{bmatrix} \sim \begin{bmatrix} 1 & 0 \\ 0 & 1 \end{bmatrix}$, $B = \begin{bmatrix} 1 & 2 \\ 1 & 2 \end{bmatrix} \sim \begin{bmatrix} 1 & 0 \\ 0 & 0 \end{bmatrix}$, and $AB = \begin{bmatrix} 5 & 10 \\ 3 & 6 \end{bmatrix} \sim \begin{bmatrix} 1 & 0 \\ 0 & 0 \end{bmatrix}$, the ranks of A, B, and AB are 2, 1, and 1, respectively.

Example 10 The rank of $\begin{bmatrix} 0 & ① & 3 & 0 & 2 \\ 0 & 0 & 0 & ① & 1 \\ 0 & 0 & 0 & 0 & 0 \end{bmatrix}$ is two and does not exceed three and five.

Finally, we show that the canonical form is unique. Suppose that $A \sim B$ and that the i-th columns of A and B are a_i and b_i, respectively. Since a linear relation that is true in A is true in B as well, if a_j is linearly independent of the vectors, so is b_j. Suppose further that B is in the canonical form. If the number of independent vectors on the left is $k - 1$, i.e., b_j is the k-th row, then b_k should be e_k, the column vector such that the k-th element is one, and the other elements are zero. Otherwise, the k-th row of the canonical form is a zero vector, or a column vector that is right from b_j becomes e_k, which contradicts that B is in the canonical form. On the other hand, if a_j can be written as $\sum_{i<j} r_i a_i$, then b_j should be written as $\sum_{i<j} r_i b_i = \sum_{i<j} r_i e_i$, which means that b_j is a column vector whose i-th element is the coefficient r_i in a_j. In any case, given A, the canonical form B is unique.

1.4 Vector Spaces and Their Dimensions

We refer to any subset V of $\mathbb{R}^n$ such that

$$\begin{cases} x, y \in V \Longrightarrow x + y \in V \\ a \in \mathbb{R}, \ x \in V \Longrightarrow ax \in V \end{cases} \tag{1.2}$$

as a linear subspace of $\mathbb{R}^n$. We may similarly define a subspace of V.

Example 11 Let V be the subset of $\mathbb{R}^n$ such that the last element of $x \in V$ is equal to the sum of the other elements. Since we can see that

$$x, y \in V \Longrightarrow x_n = \sum_{i=1}^{n-1} x_i, \ y_n = \sum_{i=1}^{n-1} y_i \Longrightarrow x_n + y_n = \sum_{i=1}^{n-1} (x_i + y_i) \Longrightarrow x + y \in V$$

and

$$x \in V \Longrightarrow x_n = \sum_{i=1}^{n-1} x_i \Longrightarrow a x_n = \sum_{i=1}^{n-1} a x_i \Longrightarrow ax \in V,$$

V satisfies (1.2) and is a subspace of $\mathbb{R}^n$. For example, for a subset W of V such that the first element is zero, $W := \{[x_1, \ldots, x_n]^T \in V | x_1 = 0\}$ satisfies (1.2), and W is a subspace of V.

In the following, we refer to any subspace of $\mathbb{R}^n$ solely as a vector space.[1] Any vector in the vector space can be expressed as a linear combination of a finite number of vectors. For example, an arbitrary $x = [x_1, x_2, x_3] \in \mathbb{R}^3$ can be written as $x = \sum_{i=1}^{3} x_i e_i$ using $e_1 := [1, 0, 0]^T$, $e_2 := [0, 1, 0]^T$, $e_3 := [0, 0, 1]^T$. We refer to a linearly independent subset $\{a_1, \ldots, a_r\}$ of V any element in V that can be expressed as a linear combination of $a_1, \ldots, a_r$ as a basis of V, and the number r of elements in $\{a_1, \ldots, a_r\}$ to as the dimension of V. Although the basis of the V is not unique, the dimension of any basis is equal. Thus, the dimension of V is unique.

Example 12 The set of vectors V that are linear combinations of $a_1, \ldots, a_5$ in Example 8 satisfies (1.2) and constitutes a vector space. Since a_1, a_2, and a_4 are linearly independent and a_3 and a_5 can be expressed by linear combinations of these vectors, each element v in V can be expressed by specifying $x_1, x_2, x_4 \in \mathbb{R}$ in $x_1 a_1 + x_2 a_2 + x_4 a_4$, but there exists a $v \in V$ that cannot be expressed by specifying $x_1, x_2 \in \mathbb{R}$ and $x_2, x_4 \in \mathbb{R}$ in $x_1 a_1 + x_2 a_2$ and $x_2 a_2 + x_4 a_4$, respectively. On the other hand, if we specify $x_1, x_2, x_3, x_4 \in \mathbb{R}$ in $x_1 a_1 + x_2 a_2 + a_3 x_3 + x_4 a_4$, then from

$$x_1 a_1 + x_2 a_2 + a_3 x_3 + x_4 a_4 = x_1 a_1 + x_2 a_2 + x_3(-a_1 + 2a_2) + x_4 a_4$$

$$= (x_1 - x_3)a_1 + (x_2 + 2x_3)a_2 + x_4 a_4,$$

there is more than one way to express $v = a_2$, such as $(x_1, x_2, x_3, x_4) = (0, 1, 0, 0)$, $(1, -1, 1, 0)$. Therefore, $\{a_1, a_2, a_4\}$ is a basis, and the dimension of the vector space is three. In addition, $\{a_1', a_2', a_4\}$, such that $a_1' = a_1 + a_2$ and $a_2' = a_1 - a_2, a_4$, is a basis as well. In fact, because of

$$[a_1', a_2', a_4] = \begin{bmatrix} 2 & 0 & -2 \\ 3 & -1 & -4 \\ 3 & 3 & 1 \\ -1 & 1 & -1 \end{bmatrix} \sim \begin{bmatrix} 1 & 0 & 0 \\ 0 & 1 & 0 \\ 0 & 0 & 1 \\ 0 & 0 & 0 \end{bmatrix},$$

they are linearly independent.

---

[1] In general, any subset V that satisfies (1.2) is said to be a vector space with scalars in $\mathbb{R}$.

Let V and W be subspaces of $\mathbb{R}^n$ and $\mathbb{R}^m$, respectively. We refer to any map $V \ni x \mapsto Ax \in W$ as the linear map[2] w.r.t. $A \in \mathbb{R}^{m \times n}$.

For example, the image $\{Ax \mid x \in V\}$ and the kernel $\{x \in V \mid Ax = 0\}$ are subspaces W and V, respectively. On the other hand, the image can be expressed as a linear combination of the columns in A and its dimension coincides with the rank of A (i.e., the number of linearly independent vectors in A).

Example 13 For the matrix A in Example 8 and vector space V, each element of which can be expressed by a linear combination of $a_1, \ldots, a_5$, the vectors in the image can be expressed by a linear combination of $a_1, a_2,$ and a_4.

$$Ax = 0 \iff \begin{bmatrix} 1 & 0 & -1 & 0 & 2 \\ 0 & 1 & 2 & 0 & -1 \\ 0 & 0 & 0 & 1 & 1 \end{bmatrix} \begin{bmatrix} x_1 \\ x_2 \\ x_3 \\ x_4 \\ x_5 \end{bmatrix} = \begin{bmatrix} 0 \\ 0 \\ 0 \end{bmatrix}$$

$$\iff \begin{bmatrix} x_1 \\ x_2 \\ x_3 \\ x_4 \\ x_5 \end{bmatrix} = \begin{bmatrix} x_3 - 2x_5 \\ -2x_3 + x_5 \\ x_3 \\ -x_5 \\ x_5 \end{bmatrix} = x_3 \begin{bmatrix} 1 \\ -2 \\ 1 \\ 0 \\ 0 \end{bmatrix} + x_5 \begin{bmatrix} -2 \\ 1 \\ 0 \\ -1 \\ 1 \end{bmatrix}.$$

The image and kernel are $\{c_1 a_1 + c_2 a_2 + c_4 a_4 \mid c_1, c_2, c_4 \in \mathbb{R}\}$ and $\left\{ c_3 \begin{bmatrix} 1 \\ -2 \\ 1 \\ 0 \\ 0 \end{bmatrix} + c_5 \begin{bmatrix} -2 \\ 1 \\ 0 \\ -1 \\ 1 \end{bmatrix} \mid c_3, c_5 \in \mathbb{R} \right\}$, respectively, and they are the subspaces of V (three dimensions) and $W = \mathbb{R}^5$ (two dimensions), respectively.

Proposition 4 *Let V and W be subspaces of $\mathbb{R}^n$ and $\mathbb{R}^m$, respectively. The image and kernel of the linear map $V \to W$ w.r.t. $A \in \mathbb{R}^{m \times n}$ are subspaces of W and V, respectively, and the sum of the dimensions is n. The dimension of the image coincides with the rank of A.*

For the proof, see the Appendix at the end of this chapter.

---

[2]In general, for vector spaces V and W, we say that $f : V \to W$ is a linear map if $f(x + y) = f(x) + f(y)$, where $x, y \in V$, $f(ax) = af(x)$, $a \in \mathbb{R}$, and $x \in V$.

1.5 Eigenvalues and Eigenvectors

For a matrix $A \in \mathbb{R}^{n \times n}$, if there exist $0 \neq x \in \mathbb{C}^n$ and $\lambda \in \mathbb{C}$ such that $Ax = \lambda x$, we refer to $x \neq 0$ as the eigenvector of eigenvalue λ. In general,

the solution of $(A - \lambda I)x = 0$ is only $x = 0 \iff \det(A - \lambda I) \neq 0$.

Combined with Proposition 1, we have the following proposition.

Proposition 5 λ *is an eigenvalue of A $\iff \det(A - \lambda I) = 0$*

In this book, we only consider matrices for which all the eigenvalues are real. In general, if the eigenvalues of $A \in \mathbb{R}^{n \times n}$ are $\lambda_1, \ldots, \lambda_n$, they are the solutions of the eigenpolynomial $\det(A - tI) = (\lambda_1 - t) \ldots (\lambda_n - t) = 0$, and if we substitute in $t = 0$, we have $\det(A) = \lambda_1 \ldots \lambda_n$.

Proposition 6 *The determinant of a square matrix is the product of its eigenvalues.*

In general, for each $\lambda \in \mathbb{R}$, the subset $V_\lambda := \{x \in \mathbb{R}^n \mid Ax = \lambda x\}$ constitutes a subspace of $\mathbb{R}^n$ (the eigenspace of λ):

$$x, y \in V_\lambda \Longrightarrow Ax = \lambda x, \ Ay = \lambda y \Longrightarrow A(x + y) = \lambda(x + y) \Longrightarrow x + y \in V_\lambda$$

$$x \in V_\lambda, \ a \in \mathbb{R} \Longrightarrow Ax = \lambda x, a \in \mathbb{R} \Longrightarrow A(ax) = \lambda(ax) \Longrightarrow ax \in V_\lambda .$$

Example 14 For $A = \begin{bmatrix} 7 & 12 & 0 \\ -2 & -3 & 0 \\ 2 & 4 & 1 \end{bmatrix}$, from $\det(A - tI) = 0$, we have $(t-1)^2(t-3) = 0$.

When $t = 1$, we have $A - tI = \begin{bmatrix} 6 & 12 & 0 \\ -2 & -4 & 0 \\ 2 & 4 & 0 \end{bmatrix} \sim \begin{bmatrix} 1 & 2 & 0 \\ 0 & 0 & 0 \\ 0 & 0 & 0 \end{bmatrix}$, and a basis

of its kernel consists of $\begin{bmatrix} 2 \\ -1 \\ 0 \end{bmatrix}$ and $\begin{bmatrix} 0 \\ 0 \\ 1 \end{bmatrix}$.

When $t = 3$, we have $A - tI = \begin{bmatrix} 4 & 12 & 0 \\ -2 & -6 & 0 \\ 2 & 4 & -2 \end{bmatrix} \sim \begin{bmatrix} 1 & 3 & 0 \\ 1 & 2 & -1 \\ 0 & 0 & 0 \end{bmatrix}$, and a

basis of its kernel consists of $\begin{bmatrix} 3 \\ -1 \\ 1 \end{bmatrix}$. Hence, we have

$$W_1 = \left\{ c_1 \begin{bmatrix} 2 \\ -1 \\ 0 \end{bmatrix} + c_2 \begin{bmatrix} 0 \\ 0 \\ 1 \end{bmatrix} \ \middle|\ c_1, c_2 \in \mathbb{R} \right\}, \quad W_3 = \left\{ c_3 \begin{bmatrix} 3 \\ -1 \\ 1 \end{bmatrix} \ \middle|\ c_3 \in \mathbb{R} \right\}.$$

Example 15 For $A = \begin{bmatrix} 1 & 3 & 2 \\ 0 & -1 & 0 \\ 1 & 2 & 0 \end{bmatrix}$, from $\det(A - tI) = 0$, we have $(t +$

$1)^2(t - 2) = 0$. When $t = -1$, we have $A - tI = \begin{bmatrix} 2 & 3 & 2 \\ 0 & 0 & 0 \\ 1 & 2 & 1 \end{bmatrix} \sim \begin{bmatrix} 1 & 0 & 1 \\ 0 & 1 & 0 \\ 0 & 0 & 0 \end{bmatrix}$,

and a basis of its kernel consists of $\begin{bmatrix} -1 \\ 0 \\ 1 \end{bmatrix}$. When $t = 2$, we have $A - tI =$

$\begin{bmatrix} -1 & 3 & 2 \\ 0 & -3 & 0 \\ 1 & 2 & -2 \end{bmatrix} \sim \begin{bmatrix} 1 & 0 & -2 \\ 0 & 1 & 0 \\ 0 & 0 & 0 \end{bmatrix}$, and a basis of its kernel consists of $\begin{bmatrix} 2 \\ 0 \\ 1 \end{bmatrix}$.

Hence, we have

$$W_{-1} = \left\{ c_1 \begin{bmatrix} -1 \\ 0 \\ 1 \end{bmatrix} \mid c_1 \in \mathbb{R} \right\}, \, W_2 = \left\{ c_2 \begin{bmatrix} 2 \\ 0 \\ 1 \end{bmatrix} \mid c_2 \in \mathbb{R} \right\}.$$

If we obtain a diagonal matrix by multiplying a square matrix $A \in \mathbb{R}^{n \times n}$ by a nonsingular matrix and its inverse from left and right, respectively, then we say that A is diagonalizable.

Example 16 If we write the matrix that arranges the eigenvectors in Example 14 as $P = \begin{bmatrix} 2 & 0 & 3 \\ -1 & 0 & -1 \\ 0 & 1 & 1 \end{bmatrix}$, then we have $P^{-1}AP = \begin{bmatrix} 1 & 0 & 0 \\ 0 & 1 & 0 \\ 0 & 0 & 3 \end{bmatrix}$.

As in Example 14, if the sum of the dimensions of the eigenspaces is n, we can diagonalize matrix A. On the other hand, as in Example 15, we cannot diagonalize A. In fact, each column vector of P should be an eigenvector. If the sum of the dimensions of the eigenspaces is less than n, we cannot choose linearly independent columns of P.

1.6 Orthonormal Bases and Orthogonal Matrix

We define the inner product and norm[3] of a vector space V as $u^T v = \sum_{i=1}^{n} u_i v_i$ and $\|u\| = \sqrt{u^T u}$, respectively, for $u, v \in V$. If a basis $u_1, \ldots, u_n$ of V is orthogonal (the inner product of each pair is zero), the norms are ones; we say that they

---

[3] In general, we say that the map $(\cdot, \cdot)$ is an inner product of V if $(u + u', v) = (u, v) + (u', v)$, $(cu, v) = c(u, v)$, $(u, v) = (u', v)$, and $u \neq 0 \implies (u, u) > 0$ for $u, v \in V$, where $c \in \mathbb{R}$.

constitute an orthonormal basis. For an arbitrary linear independent $v_1, \ldots, v_n \in V$, we construct an orthonormal basis $u_1, \ldots, u_n$ of V such that the subspaces that contain $u_1, \ldots, u_i$ and $v_1, \ldots, v_i$ coincide for $i = 1, \ldots, n$.

Example 17 (Gram–Schmidt Orthonormal Basis) We construct an orthonormal basis $u_1, \ldots, u_i$ such that

$$\{\alpha_1 v_1 + \ldots + \alpha_i v_i \,|\, \alpha_1, \ldots, \alpha_i \in \mathbb{R}\} = \{\beta_1 u_1 + \ldots + \beta_i u_i \,|\, \beta_1, \ldots, \beta_i \in \mathbb{R}\}$$

for each $i = 1, \ldots, n$: Suppose we are given $v_1 = \begin{bmatrix} 1 \\ 1 \\ 0 \end{bmatrix}$, $v_2 = \begin{bmatrix} 1 \\ 3 \\ 1 \end{bmatrix}$, and $v_3 =$

$\begin{bmatrix} 2 \\ -1 \\ 1 \end{bmatrix}$. Then, the orthonormal basis consists of $u_1 = \dfrac{1}{\|v_1\|} = \dfrac{1}{\sqrt{2}} \begin{bmatrix} 1 \\ 1 \\ 0 \end{bmatrix}$,

$$v_2' = v_2 - (v_2, u_1)u_1 = \begin{bmatrix} 1 \\ 3 \\ 1 \end{bmatrix} - \frac{4}{\sqrt{2}} \cdot \frac{1}{\sqrt{2}} \begin{bmatrix} 1 \\ 1 \\ 0 \end{bmatrix} = \begin{bmatrix} -1 \\ 1 \\ 1 \end{bmatrix}, \quad u_2 = \frac{v_2'}{\|v_2'\|} =$$

$\dfrac{1}{\sqrt{3}} \begin{bmatrix} -1 \\ 1 \\ 1 \end{bmatrix}$

$$v_3' = v_3 - (v_3, u_1)u_1 - (v_3, u_2)u_2 = \begin{bmatrix} 2 \\ -1 \\ 1 \end{bmatrix} - \frac{1}{\sqrt{2}} \cdot \frac{1}{\sqrt{2}} \begin{bmatrix} 1 \\ 1 \\ 0 \end{bmatrix} - \frac{-2}{\sqrt{3}} \cdot \frac{1}{\sqrt{3}} \begin{bmatrix} -1 \\ 1 \\ 1 \end{bmatrix} =$$

$\dfrac{5}{6} \begin{bmatrix} 1 \\ -1 \\ 2 \end{bmatrix}$, and $u_3 = \dfrac{1}{\sqrt{6}} \begin{bmatrix} 1 \\ -1 \\ 2 \end{bmatrix}$.

We say a square matrix such that the columns are orthonormal. For an orthogonal matrix $P \in \mathbb{R}^{n \times n}$, $P^T P$ is the unit matrix. Therefore,

$$P^T = P^{-1}. \tag{1.3}$$

If we take the determinants on both sides, then $\det(P^T)\det(P) = 1$. From $\det(P) = \det(P^T)$, we have $\det(P) = \pm 1$. On the other hand, we refer to the linear map $V \ni x \mapsto Px \in V$ as an orthogonal map. Since $(Px)^T (Py) = x^T P^T P y = x^T y$, $x, y \in V$, an orthogonal map does not change the inner product of any pairs in V.

1.7 Diagonalization of Symmetric Matrices

In this section, we assume that square matrix $A \in \mathbb{R}^{n \times n}$ is symmetric.

We say that a square matrix is an upper-triangular matrix if the (i, j)-th elements are zero for all $i > j$. Then, we have the following proposition.

Proposition 7 *For any square matrix A, we can obtain an upper-triangular matrix $P^{-1}AP$ by multiplying it from right using an orthogonal matrix P.*

For the proof, see the Appendix at the end of this chapter.

If we note that $P^{-1}AP$ is symmetric because $(P^{-1}AP)^T = P^T A^T (P^{-1})^T = P^{-1}AP$, from (1.3), Proposition 7 claims that diagonalization and triangulation are obtained using the orthogonal matrix P.

In the following, we claim a stronger statement. To this end, we note the following proposition.

Proposition 8 *For a symmetric matrix, any eigenvectors in different eigenspaces are orthogonal.*

In fact, for eigenvalues $\lambda, \mu \in \mathbb{R}$ of A, where $x \in V_\lambda$ and $y \in V_\mu$, we have

$$\lambda x^T y = (\lambda x)^T y = (Ax)^T y = x^T A^T y = x^T A y = x^T (\mu y) = \mu x^T y .$$

In addition, because $\lambda \neq \mu$, we have $x^T y = 0$. As we have seen before, a matrix A being diagonalizable is equivalent to the sum n of the dimensions of the eigenspaces. Thus, if we choose the basis of each eigenspace to be orthogonal, all n vectors will be orthogonal.

Proposition 9 *For a symmetric matrix A and an orthogonal matrix P, the $P^{-1}AP$ is diagonal with diagonal elements equal to the eigenvalues of A.*

Example 18 The eigenspaces of $\begin{bmatrix} 1 & 2 & -1 \\ 2 & -2 & 2 \\ -1 & 2 & 1 \end{bmatrix}$ are $\left\{ c_1 \begin{bmatrix} 2 \\ 1 \\ 0 \end{bmatrix} + c_2 \begin{bmatrix} -1 \\ 0 \\ 1 \end{bmatrix} \right.$

$\left. | \ c_1, c_2 \in \mathbb{R} \right\}$ and $\left\{ c_3 \begin{bmatrix} 1 \\ -2 \\ 1 \end{bmatrix} \ | \ c_3 \in \mathbb{R} \right\}$.

Then, we orthogonalize the basis of the two-dimensional eigenspace. For $P = \begin{bmatrix} 2/\sqrt{5} & -1/\sqrt{30} & 1/\sqrt{6} \\ 1/\sqrt{5} & 2/\sqrt{30} & -2/\sqrt{6} \\ 0 & 5/\sqrt{30} & 1/\sqrt{6} \end{bmatrix}$, we have $P^{-1}AP = \begin{bmatrix} 2 & 0 & 0 \\ 0 & 2 & 0 \\ 0 & 0 & -4 \end{bmatrix}$.

In addition, from the discussion thus far, we have the following proposition.

Proposition 10 *For a symmetric matrix A of size n, the three conditions below are equivalent:*

1. *A matrix $B \in \mathbb{R}^{m \times n}$ exists such that $A = B^T B$.*
2. *$x^T A x \geq 0$ for an arbitrary $x \in \mathbb{R}^n$.*
3. *All the eigenvalues of A are nonnegative.*

In fact, 1. $\implies$ 2. because $A = B^T B \implies x^T A x = x^T B^T B x = \|Bx\|^2$, 2. $\implies$ 3. because $x^T A x \geq 0 \implies 0 \leq x^T A x = x^T \lambda x = \lambda \|x\|^2$, and 3. $\implies$ 1. because $\lambda_1, \ldots, \lambda_n \geq 0 \implies A = P^{-1}DP = P^T \sqrt{D}\sqrt{D}P = (\sqrt{D}P)^T \sqrt{D}P$, where D and $\sqrt{D}$ are the diagonal matrices whose elements are $\lambda_1, \ldots, \lambda_n$ and $\sqrt{\lambda_1}, \ldots, \sqrt{\lambda_n}$, respectively.

In this book, we refer to the matrices that satisfy the three equivalent conditions in Proposition 10 and the ones whose eigenvalues are positive as to nonnegative definite and positive definite matrices, respectively.

Appendix: Proof of Propositions

Proposition 2 *For square matrices A and B of the same size, we have $\det(AB) = \det(A)\det(B)$ and $\det(A^T) = \det(A)$.*

Proof For Steps 1, 2, and 3, we multiply the following matrices from left:

$V_i(\alpha)$:   a unit matrix where the (i, i)-th element has been replaced with α.

$U_{i,j}$:   a unit matrix where the (i, i), (j, j)-th and (i, j), (j, i)-th elements have been replaced by zero and one, respectively.

$W_{i,j}(\beta)$:   a unit matrix where the (i, j)-th zero $(i \neq j)$ has been replaced by $-\beta$.

Then, for $B \in \mathbb{R}^{n \times n}$,

$$\det(V_i(\alpha)B) = \alpha\det(B), \ \det(U_{i,j}B) = -\det(B), \ \det(W_{i,j}(\beta)B) = \det(B) .$$
$$(1.4)$$

Since

$$\det(V_i(\alpha)) = \alpha, \ \det(U_{i,j}) = -1, \ \det(W_{i,j}(\beta)) = 1 \qquad (1.5)$$

holds, if we write matrix A as the product $E_1, \ldots, E_r$ of matrices of the three types, then we have

$$\det(A) = \det(E_1) \ldots \det(E_r) .$$

$$\det(AB) = \det(E_1 \cdot E_2 \ldots E_r B) = \det(E_1)\det(E_2 \ldots E_r B) = \ldots$$
$$= \det(E_1) \ldots \det(E_r)\det(B) = \det(A)\det(B).$$

On the other hand, since matrices $V_i(\alpha)$ and $U_{i,j}$ are symmetric and $W_{i,j}(\beta)^T = W_{j,i}(\beta)$, we have a similar equation to (1.4) and (1.5). Hence, we have

$$\det(A^T) = \det(E_r^T \ldots E_1^T) = \det(E_r^T) \ldots \det(E_1^T) = \det(E_1) \ldots \det(E_r) = \det(A) .$$

$\square$

Proposition 4 *Let V and W be subspaces of $\mathbb{R}^n$ and $\mathbb{R}^m$, respectively. The image and kernel of the linear map $V \to W$ w.r.t. $A \in \mathbb{R}^{m \times n}$ are subspaces of W and V, respectively, and the sum of the dimensions is n. The dimension of the image coincides with the rank of A.*

Proof Let r and $x_1, \ldots, x_r \in V$ be the dimension and basis of the kernel, respectively. We add $x_{r+1}, \ldots, x_n$, which are linearly independent of them, so that $x_1, \ldots, x_r, x_{r+1}, \ldots, x_n$ are the bases of V. It is sufficient to show that $Ax_{r+1}, \ldots, Ax_n$ are the bases of the image.

First, since $x_1, \ldots, x_r$ are vectors in the kernel, we have $Ax_1 = \ldots = Ax_r = 0$. For an arbitrary $x = \sum_{j=1}^n b_j x_j$ with $b_{r+1}, \ldots, b_n \in \mathbb{R}$, the image can be expressed as $Ax = \sum_{j=r+1}^n b_j Ax_j$, which is a linear combination of $Ax_{r+1}, \ldots, Ax_n$. Then, our goal is to show that

$$\sum_{i=r+1}^n b_i Ax_i = 0 \Longrightarrow b_{r+1}, \ldots, b_n = 0. \tag{1.6}$$

If $A \sum_{i=r+1}^n b_i x_i = 0$, then $\sum_{i=r+1}^n b_i x_i$ is in the kernel. Therefore, there exist $b_1, \ldots, b_r$ such that $\sum_{i=r+1}^n b_i x_i = -\sum_{i=1}^r b_i x_i$, which means that $\sum_{i=1}^n b_i x_i = 0$. However, we assumed that $x_1, \ldots, x_n$ are linearly independent, which means that $b_1 = \ldots = b_n = 0$, and Proposition (1.6) is proven. □

Proposition 7 *For any square matrix A, we can obtain an upper-triangular matrix $P^{-1}AP$ by multiplying it, using an orthonormal matrix P.*

Proof We prove the proposition by induction. For $n = 1$, since the matrix is scalar, the claim holds. From the assumption of induction, for an arbitrary $\tilde{B} \in \mathbb{R}^{(n-1) \times (n-1)}$, there exists an orthogonal matrix $\tilde{Q}$ such that

$$\tilde{Q}^{-1} \tilde{B} \tilde{Q} = \begin{bmatrix} \tilde{\lambda}_2 & & * \\ & \ddots & \\ 0 & & \tilde{\lambda}_n \end{bmatrix},$$

where $*$ represents the nonzero elements and $\tilde{\lambda}_2, \ldots, \tilde{\lambda}_n$ are the eigenvalues of $\tilde{B}$.

For a nonsingular matrix $A \in \mathbb{R}^{n \times n}$ with eigenvalues $\lambda_1, \ldots, \lambda_n$, allowing multiplicity, let u_1 be an eigenvector of eigenvalue λ_1 and R an orthogonal matrix such that the first column is u_1. Then, we have $Re_1 = u_1$ and $Au_1 = \lambda_1 u_1$, where $e_1 := [1, 0, \ldots, 0]^T \in \mathbb{R}^n$. Hence, we have

$$R^{-1}ARe_1 = R^{-1}Au_1 = \lambda_1 R^{-1}u_1 = \lambda_1 R^{-1}Re_1 = \lambda_1 e_1,$$

and we may express

$$R^{-1}AR = \begin{bmatrix} \lambda_1 & b \\ 0 & B \end{bmatrix},$$

where $b \in \mathbb{R}[1 \times (n-1)]$ and $0 \in \mathbb{R}^{(n-1)\times 1}$. Note that R and A are nonsingular, so is B.

We claim that $P = R \begin{bmatrix} 1 & 0 \\ 0 & Q \end{bmatrix}$ is an orthogonal matrix, where Q is an orthogonal matrix that diagonalizes $B \in \mathbb{R}^{(n-1)\times(n-1)}$. In fact, $Q^T Q$ is a unit matrix, so is $P^T P = \begin{bmatrix} 1 & 0 \\ 0 & Q \end{bmatrix} R^T R \begin{bmatrix} 1 & 0 \\ 0 & Q \end{bmatrix}$. Note that the eigenvalues of B are $\lambda_2, \ldots, \lambda_n$ of A:

$$\prod_{i=1}^{n} (\lambda_i - \lambda) = \det(A - \lambda I_n) = \det(R^{-1} A R - \lambda I_n) = (\lambda_1 - \lambda) \det(B - \lambda I_{n-1}),$$

where I_n is a unit matrix of size n.

Finally, we claim that A is diagonalized by multiplying P^{-1} and P from left and right, respectively:

$$P^{-1} A P = \begin{bmatrix} 1 & 0 \\ 0 & Q^{-1} \end{bmatrix} R^{-1} A R \begin{bmatrix} 1 & 0 \\ 0 & Q \end{bmatrix} = \begin{bmatrix} 1 & 0 \\ 0 & Q^{-1} \end{bmatrix} \begin{bmatrix} \lambda_1 & b \\ 0 & B \end{bmatrix} \begin{bmatrix} 1 & 0 \\ 0 & Q \end{bmatrix}$$

$$= \begin{bmatrix} \lambda_1 & bQ \\ 0 & Q^{-1} B Q \end{bmatrix} = \begin{bmatrix} \lambda_1 & & & * \\ & \lambda_2 & & \\ & & \ddots & \\ & & & \lambda_n \end{bmatrix},$$

which completes the proof. $\qquad \square$

Chapter 2
Linear Regression

Abstract Fitting covariate and response data to a line is referred to as linear regression. In this chapter, we introduce the least squares method for a single covariate (single regression) first and extend it to multiple covariates (multiple regression) later. Then, based on the statistical notion of estimating parameters from data, we find the distribution of the coefficients (estimates) obtained via the least squares method. Thus, we present a method for estimating a confidence interval of the estimates and for testing whether each of the true coefficients is zero. Moreover, we present a method for finding redundant covariates that may be removed. Finally, we consider obtaining a confidence interval of the response of new data outside of the dataset used for the estimation. The problem of linear regression is a basis of consideration in various issues and plays a significant role in machine learning. .

2.1 Least Squares Method

Let N be a positive integer. For given data $(x_1, y_1), \ldots, (x_N, y_N) \in \mathbb{R} \times \mathbb{R}$, we obtain the intercept β_0 and slope β_1 via the least squares method. More precisely, we minimize the sum $L := \sum_{i=1}^{N} (y_i - \beta_0 - \beta_1 x_i)^2$ of the squared distances $(y_i - \beta_0 - \beta_1 x_i)^2$ between (x_i, y_i) and $(x_i, \beta_0 + x_i \beta_1)$ over $i = 1, \cdots, N$ (Fig. 2.1).

Then, by partially differentiating L by β_0, β_1 and letting them be zero, we obtain the following equations:

$$\frac{\partial L}{\partial \beta_0} = -2 \sum_{i=1}^{N} (y_i - \beta_0 - \beta_1 x_i) = 0 \tag{2.1}$$

$$\frac{\partial L}{\partial \beta_1} = -2 \sum_{i=1}^{N} x_i (y_i - \beta_0 - \beta_1 x_i) = 0, \tag{2.2}$$

where the partial derivative is calculated by differentiating each variable and regarding the other variables as constants. In this case, β_0 and β_1 are regarded as constants when differentiating L by β_1 and β_0, respectively.

© The Author(s), under exclusive license to Springer Nature Singapore Pte Ltd. 2021
J. Suzuki, *Statistical Learning with Math and Python*,
https://doi.org/10.1007/978-981-15-7877-9_2

Fig. 2.1 Obtain β_0 and β_1 that minimize $\sum_{i=1}^{n}(y_i - \beta_1 x_i - \beta_0)^2$ via the least squares method

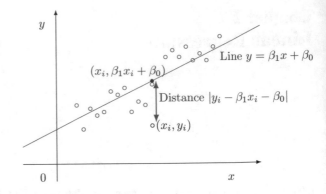

By solving Eqs. (2.1)–(2.2) when $\sum_{i=1}^{N}(x_i - \bar{x})^2 \neq 0$, i.e.,

$$x_1 = \cdots = x_N \text{ is not true} \tag{2.3}$$

we obtain

$$\hat{\beta}_1 = \frac{\sum\limits_{i=1}^{N}(x_i - \bar{x})(y_i - \bar{y})}{\sum\limits_{i=1}^{N}(x_i - \bar{x})^2} \tag{2.4}$$

$$\hat{\beta}_0 = \bar{y} - \hat{\beta}_1 \bar{x} , \tag{2.5}$$

where $\bar{x} := \dfrac{1}{N}\sum\limits_{i=1}^{N} x_i$ and $\bar{y} := \dfrac{1}{N}\sum\limits_{i=1}^{N} y_i$. We used the variables $\hat{\beta}_0$ and $\hat{\beta}_1$ instead of β_0 and β_1, which means that they are not the true values but rather estimates obtained from data.

If we divide both sides of Eq. (2.1) by $-2N$, we obtain (2.5). To show (2.4), we center the data as follows:

$$\tilde{x}_1 := x_1 - \bar{x}, \ldots, \tilde{x}_N := x_N - \bar{x}, \tilde{y}_1 := y_1 - \bar{y}, \ldots, \tilde{y}_N := y_N - \bar{y},$$

and obtain the slope ($\hat{\beta}_1$) first. Even if we shift all the points by $(\bar{x}, \bar{y})$ in the directions of X and Y, the slope remains the same, but the line goes through the origin. Note that once $x_1, \ldots, x_N, y_1, \ldots, y_N$ are centered, then we have

$$\frac{1}{N}\sum_{i=1}^{N} \tilde{x}_i = \frac{1}{N}\sum_{i=1}^{N} \tilde{y}_i = 0$$

Fig. 2.2 Instead of (2.4), we
center the data at the
beginning and obtain the
slope via (2.6) first and obtain
the intercept later via the
arithmetic means $\bar{x}, \bar{y}$ and the
relation in (2.5)

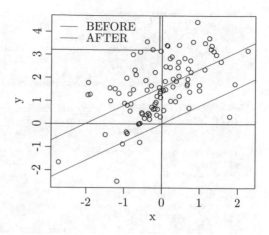

and

$$\frac{1}{N}\sum_{i=1}^{N}\tilde{y}_i - \beta_1 \frac{1}{N}\sum_{i=1}^{N}\tilde{x}_i = 0 \, ,$$

which means that the intercept becomes zero with the new coordinates. From the
centered $x_1, \ldots, x_N$ and $y_1, \ldots, y_N$, we obtain $\hat{\beta}_1$; if we substitute $\beta_0 = 0$ into
(2.2), unless $\tilde{x}_1 = \cdots = \tilde{x}_N = 0$, we obtain

$$\hat{\beta}_1 = \frac{\displaystyle\sum_{i=1}^{N}\tilde{x}_i\tilde{y}_i}{\displaystyle\sum_{i=1}^{N}\tilde{x}_i^2} \, . \tag{2.6}$$

The estimate (2.6) is obtained after centering w.r.t. $x_1, \ldots, x_N$ and $y_1, \ldots, y_N$, and
if we return to the values before the centering by

$$x_1 := \tilde{x}_1 + \bar{x}, \ldots, x_N := \tilde{x}_N + \bar{x}, \, y_1 := \tilde{y}_1 + \bar{y}, \ldots, y_N := \tilde{y}_N + \bar{y} \, ,$$

we obtain (2.4). Finally, from $\hat{\beta}_1$ and the relation in (2.5), we obtain the intercept
$\hat{\beta}_0 = \bar{y} - \hat{\beta}_1\bar{x}$.

Example 19 Figure 2.2 shows the two lines l and l' generated via the Python
program below. l is obtained from the N pairs of data and the least squares method,
and l' is obtained by shifting l so that it goes through the origin.

```
def min_sq(x,y):
    x_bar,y_bar=np.mean(x),np.mean(y)
    beta_1=np.dot(x-x_bar,y-y_bar)/np.linalg.norm(x-x_bar)**2
    beta_0=y_bar-beta_1*x_bar
    return [beta_1,beta_0]
```

```
N=100             # Data generation
a=np.random.normal(loc=2,scale=1,size=N)  # randomly generate the
   coefficients of the line
b=randn(1) # randomly generate the points surrounding the line
x=randn(N)
y=a*x+b+randn(N)

a1,b1=min_sq(x,y)   # estimating coefficients
xx=x-np.mean(x);yy=y-np.mean(y)  #  centering
a2,b2=min_sq(xx,yy)
```

```
(1.7865393047324676, 1.067565008452225e-16)
```

```
x_seq=np.arange(-5,5,0.1)
y_pre=x_seq*a1+b1
yy_pre=x_seq*a2+b2

plt.scatter(x,y,c="black")   # plots of the points
plt.axhline(y=0,c="black",linewidth=0.5)
plt.axvline(x=0,c="black",linewidth=0.5)
plt.plot(x_seq,y_pre,c="blue",label="Before")     # the line before centering
plt.plot(x_seq,yy_pre,c="orange",label="After")  # the line after centering
plt.legend(loc="upper_left")
```

In the program, by using x_seq with intercept a and slope b , we get y_pre and yy_pre , and the X-axis $y = 0$, and the Y-axis $x = 0$ by the commands plt.axvline(x=0), plt.axhline(y=0), and abline(v=0), respectively. The function min_sq defined in the program returns the intercept b and slope a from the least squares methods.

2.2 Multiple Regression

We extend the regression problem for a single covariate ($p = 1$) to the one for multiple covariates ($p \geq 1$). To this end, we formulate the least squares method for single regression with matrices. If we define

$$y := \begin{bmatrix} y_1 \\ \vdots \\ y_N \end{bmatrix}, \quad X := \begin{bmatrix} 1 & x_1 \\ \vdots & \vdots \\ 1 & x_N \end{bmatrix}, \quad \beta := \begin{bmatrix} \beta_0 \\ \beta_1 \end{bmatrix}, \tag{2.7}$$

then for $L := \sum_{i=1}^{N} (y_i - \beta_0 - x_i\beta_1)^2$, we have

$$L = \|y - X\beta\|^2$$

and

$$\nabla L := \begin{bmatrix} \dfrac{\partial L}{\partial \beta_0} \\ \dfrac{\partial L}{\partial \beta_1} \end{bmatrix} = -2X^T(y - X\beta) . \tag{2.8}$$

By examining (2.8), we see that the elements on the right-hand side of (2.8) are

$$\begin{bmatrix} -2\sum_{i=1}^{N}(y_i - \beta_0 - \beta_1 x_i) \\ -2\sum_{i=1}^{N} x_i(y_i - \beta_0 - \beta_1 x_i) \end{bmatrix}, \tag{2.9}$$

which means that (2.9) expresses (2.1) and (2.2).

For multiple regression ($p \geq 1$), we may extend the formulation in (2.7) to the one below:

$$y := \begin{bmatrix} y_1 \\ \vdots \\ y_N \end{bmatrix}, \ X := \begin{bmatrix} 1 & x_{1,1} & \cdots & x_{1,p} \\ \vdots & \vdots & \ddots & \vdots \\ 1 & x_{N,1} & \cdots & x_{N,p} \end{bmatrix}, \ \beta := \begin{bmatrix} \beta_0 \\ \beta_1 \\ \vdots \\ \beta_p \end{bmatrix},$$

Even if we extend this formulation, (2.8) still holds. In fact, if we let $x_{i,0} = 1$, $i = 1, \ldots, N$, (2.9) is extended to

$$-2X^T(y - X\beta) = \begin{bmatrix} -2\sum_{i=1}^{N}(y_i - \sum_{j=0}^{p}\beta_j x_{i,j}) \\ -2\sum_{i=1}^{N} x_{i,1}(y_i - \sum_{j=0}^{p}\beta_j x_{i,j}) \\ \vdots \\ -2\sum_{i=1}^{N} x_{i,p}(y_i - \sum_{j=0}^{p}\beta_j x_{i,j}) \end{bmatrix}. \tag{2.10}$$

Since (2.10) being zero means that $X^T X\beta = X^T y$, we have the following statement:

Proposition 11 *When a matrix $X^T X \in \mathbb{R}^{(p+1) \times (p+1)}$ is invertible, we have*

$$\hat{\beta} = (X^T X)^{-1} X^T y . \tag{2.11}$$

Example 20 The following is a Python program that estimates the intercept and slope via the least squares method: based on (2.11) for $N = 100$ random data points with $\beta_0 = 1$, $\beta_1 = 2$, $\beta_2 = 3$, and $\epsilon_i \sim N(0, 1)$.

```
n=100; p=2
beta=np.array([1,2,3])
x=randn(n,2)
y=beta[0]+beta[1]*x[:,0]+beta[2]*x[:,1]+randn(n)
X=np.insert(x,0,1,axis=1)  # adding the all one vector in the leftmost
    column
np.linalg.inv(X.T@X)@X.T@y  # estimate the beta
```

```
array([1.08456582, 1.91382258, 2.98813678])
```

We may notice that the matrix $X^T X$ is not invertible under each of the following conditions:

1. $N < p + 1$.
2. Two columns in X coincide.

In fact, when $N < p + 1$, from Proposition 3, we have

$$\text{rank}(X^T X) \le \text{rank}(X) \le \min\{N, p + 1\} = N < p + 1 ,$$

which means from Proposition 1 that $X^T X$ is not invertible. On the other hand, when two columns in X coincide, from Proposition 3, we have

$$\text{rank}(X^T X) \le \text{rank}(X) < p + 1 ,$$

which means that, from Proposition 1, $X^T X$ is not invertible as well.

Moreover, we see that the ranks of $X^T X$ and X coincide. In fact, for an arbitrary $z \in \mathbb{R}^{p+1}$,

$$X^T X z = 0 \Longrightarrow z^T X^T X z = 0 \Longrightarrow \|Xz\|^2 = 0 \Longrightarrow Xz = 0$$

and

$$Xz = 0 \Longrightarrow X^T X z = 0,$$

which means that the kernels of $X^T X$ and X coincide. Since the numbers of columns of $X^T X$ and X are equal, so are the dimensions of their images (see Proposition 4).

On the other hand, from Proposition 4, since the image dimensions are the ranks of the matrices, the ranks of $X^T X$ and X^T are equal.

In the following section, we assume that the rank of $X \in \mathbb{R}^{N \times (p+1)}$ is $p + 1$. In particular, if $p = 1$, the condition in (2.3) is equivalent to $\text{rank}(X) = 1 < 2 = p+1$.

2.3 Distribution of $\hat{\beta}$

We assume that the responses $y \in \mathbb{R}^N$ have been obtained from the covariates $X \in \mathbb{R}^{N \times (p+1)}$ multiplied by the (true) coefficients $\beta \in \mathbb{R}^{p+1}$ plus some noise $\epsilon \in \mathbb{R}^N$, which means that y fluctuates only because of the randomness in ϵ. Thus, we let

$$y = X\beta + \epsilon , \tag{2.12}$$

where the true β is unknown and different from the estimate $\hat{\beta}$. We have estimated $\hat{\beta}$ via the least squares method from the N pairs of data $(x_1, y_1), \ldots, (x_N, y_N) \in \mathbb{R}^p \times \mathbb{R}$, where $x_i \in \mathbb{R}^p$ is the row vector consisting of p values excluding the leftmost one in the i-th row of X.

Moreover, we assume that each element $\epsilon_1, \ldots, \epsilon_N$ in the random variable ϵ is independent of the others and follows the Gaussian distribution with mean zero and variance σ^2. Therefore, the density function is

$$f_i(\epsilon_i) = \frac{1}{\sqrt{2\pi\sigma^2}} e^{-\frac{\epsilon_i^2}{2\sigma^2}}$$

for $i = 1, \ldots, N$, which we write as $\epsilon_i \sim N(0, \sigma^2)$. We may express the distributions of $\epsilon_1, \ldots, \epsilon_N$ by

$$f(\epsilon) = \prod_{i=1}^{N} f_i(\epsilon_i) = \frac{1}{(2\pi\sigma^2)^{N/2}} e^{-\frac{\epsilon^T \epsilon}{2\sigma^2}} ,$$

which we write as $\epsilon \sim N(0, \sigma^2 I)$, where I is a unit matrix of size N.

In general, we have the following statement:

Proposition 12 *Two Gaussian random variables are independent if and only if their covariance is zero.*

For the proof, see the Appendix at the end of this chapter.

If we substitute (2.12) into (2.11), we have

$$\hat{\beta} = (X^T X)^{-1} X^T (X\beta + \epsilon) = \beta + (X^T X)^{-1} X^T \epsilon . \tag{2.13}$$

The estimate $\hat{\beta}$ of β depends on the value of ϵ because N pairs of data $(x_1, y_1), \ldots, (x_N, y_N)$ randomly occur. In fact, for the same $x_1, \ldots, x_N$, if we again generate $(y_1, \ldots, y_N)$ randomly according to (2.12) only, the fluctuations $(\epsilon_1, \ldots, \epsilon_N)$ are different. The estimate $\hat{\beta}$ is obtained based on the N pairs of randomly generated data points. On the other hand, since the average of $\epsilon \in \mathbb{R}^N$ is zero, the average of ϵ multiplied from left by the constant matrix $(X^T X)^{-1} X^T$ is zero. Therefore, from (2.13), we have

$$E[\hat{\beta}] = \beta. \tag{2.14}$$

In general, we say that an estimate is unbiased if its average coincides with the true value.

Moreover, both $\hat{\beta}$ and its average β consist of $p + 1$ values. In this case, in addition to each variance $V(\hat{\beta}_i) = E(\hat{\beta}_i - \beta_i)^2$, $i = 0, 1, \ldots, p$, the covariance $\sigma_{i,j} := E(\hat{\beta}_i - \beta_i)(\hat{\beta}_j - \beta_j)$ can be defined for each pair $i \neq j$. We refer to the matrix consisting of $\sigma_{i,j}$ in the i-th row and j-th column as to the covariance matrix of $\hat{\beta}$, which can be computed as follows. From (2.13), we have

$$E \begin{bmatrix} (\hat{\beta}_0 - \beta_0)^2 & (\hat{\beta}_0 - \beta_0)(\hat{\beta}_1 - \beta_1) & \cdots & (\hat{\beta}_0 - \beta_0)(\hat{\beta}_p - \beta_p) \\ (\hat{\beta}_1 - \beta_1)(\hat{\beta}_0 - \beta_0) & (\hat{\beta}_1 - \beta_1)^2 & \cdots & (\hat{\beta}_1 - \beta_1)(\hat{\beta}_p - \beta_p) \\ \vdots & \vdots & \ddots & \vdots \\ (\hat{\beta}_p - \beta_p)(\hat{\beta}_0 - \beta_0) & (\hat{\beta}_p - \beta_p)(\hat{\beta}_1 - \beta_1) & \cdots & (\hat{\beta}_p - \beta_p)^2 \end{bmatrix}$$

$$= E \begin{bmatrix} \hat{\beta}_0 - \beta_0 \\ \hat{\beta}_1 - \beta_1 \\ \vdots \\ \hat{\beta}_p - \beta_p \end{bmatrix} [\hat{\beta}_0 - \beta_0, \hat{\beta}_1 - \beta_1, \ldots, \hat{\beta}_p - \beta_p]$$

$$= E(\hat{\beta} - \beta)(\hat{\beta} - \beta)^T = E(X^T X)^{-1} X^T \epsilon \{(X^T X)^{-1} X^T \epsilon\}^T$$

$$= (X^T X)^{-1} X^T E \epsilon \epsilon^T X (X^T X)^{-1} = \sigma^2 (X^T X)^{-1} ,$$

for which we have determined that the covariance matrix of ϵ is $E\epsilon\epsilon^T = \sigma^2 I$. Hence, we have

$$\hat{\beta} \sim N(\beta, \sigma^2 (X^T X)^{-1}) . \tag{2.15}$$

2.4   Distribution of the RSS Values

In this subsection, we derive the distribution of the squared loss by substituting $\beta = \hat{\beta}$ into $L = \|y - X\beta\|^2$ when we fit the data to a line. To this end, we explore the properties of the matrix[1] $H := X(X^T X)^{-1} X^T \in \mathbb{R}^{N \times N}$. The following are easy to derive but useful in the later part of this book:

$$H^2 = X(X^T X)^{-1} X^T \cdot X(X^T X)^{-1} X^T = X(X^T X)^{-1} X^T = H$$

$$(I - H)^2 = I - 2H + H^2 = I - H$$

$$HX = X(X^T X)^{-1} X^T \cdot X = X.$$

---

[1] We often refer to this matrix as the hat matrix.

Moreover, if we set $\hat{y} := X\hat{\beta}$, then from (2.11), we have $\hat{y} = X\hat{\beta} = X(X^T X)^{-1}X^T y = Hy$, and

$$y - \hat{y} = (I - H)y = (I - H)(X\beta + \epsilon) = (X - HX)\beta + (1 - H)\epsilon = (I - H)\epsilon.$$

(2.16)

We define

$$RSS := \|y - \hat{y}\|^2 = \{(I - H)\epsilon\}^T (I - H)\epsilon = \epsilon^T (I - H)^2 \epsilon = \epsilon^T (I - H)\epsilon .$$

(2.17)

The following proposition is useful for deriving the distribution of the RSS values:

Proposition 13 The eigenvalues of H and $I - H$ are only zeros and ones, and the dimensions of the eigenspaces of H and $I - H$ with eigenvalues one and zero, respectively, are both $p+1$, while the dimensions of the eigenspaces of H and $I - H$ with eigenvalues of zero and one, respectively, are both $N - p - 1$.

For the proof, see the Appendix at the end of this chapter.

Since $I - H$ is a real symmetric matrix, from Proposition 9, we can diagonalize it by an orthogonal matrix P to obtain the diagonal matrix $P(I - H)P^T$. Additionally, since the $N - p - 1$ and $p + 1$ eigenvalues out of the N eigenvalues are ones and zeros, respectively, without loss of generality, we may put ones for the first $N - p - 1$ elements in the diagonal matrix:

$$P(I - H)P^T = \mathrm{diag}(\underbrace{1, \ldots, 1}_{N-p-1}, \underbrace{0, \ldots, 0}_{p+1}) .$$

Thus, if we define $v = P\epsilon \in \mathbb{R}^N$, then from $\epsilon = P^T v$ and (2.17), we have

$$RSS = \epsilon^T (I - H)\epsilon = (P^T v)^T (I - H)P^T v = v^T P(I - H)P^T v$$

$$= [v_1, \ldots, v_{N-p-1}, v_{N-p}, \ldots, v_n] \begin{bmatrix} 1 & 0 & \cdots & \cdots & \cdots & 0 \\ 0 & \ddots & 0 & \cdots & \cdots & \vdots \\ \vdots & 0 & 1 & 0 & \cdots & 0 \\ \vdots & \vdots & 0 & 0 & \cdots & \vdots \\ \vdots & \vdots & \vdots & \vdots & \ddots & \vdots \\ 0 & \cdots & 0 & \cdots & \cdots & 0 \end{bmatrix} \begin{bmatrix} v_1 \\ \vdots \\ v_{N-p-1} \\ v_{N-p} \\ \vdots \\ v_N \end{bmatrix}$$

$$= \sum_{i=1}^{N-p-1} v_i^2$$

for $v = [v_1, \ldots, v_N]^T$.

Let $w \in \mathbb{R}^{N-p-1}$ be the first $N - p - 1$ elements of v. Then, since the average of v is $E[P\epsilon] = 0$, we have $E[w] = 0$; thus,

$$Evv^T = EP\epsilon(P\epsilon)^T = PE\epsilon\epsilon^T P^T = P\sigma^2 \tilde{I} P^T = \sigma^2 \tilde{I},$$

where $\tilde{I}$ is a diagonal matrix such that the first $N - p - 1$ and last $p + 1$ diagonal elements are ones and zeros, respectively. Hence, the covariance matrix is $Eww^T = \sigma^2 I$, where I is a unit matrix of size $N - p - 1$.

For the Gaussian distributions, the independence of variables is equivalent to the covariance matrix being a diagonal matrix (Proposition 12); we have

$$\frac{RSS}{\sigma^2} \sim \chi^2_{N-p-1}, \tag{2.18}$$

where we denote by χ^2_m, which is a χ^2 distribution with m degrees of freedom, i.e., the distribution of the squared sum of m independent standard Gaussian random variables.

Example 21 For each degree of freedom up to m for the χ^2 distribution, we depict the probability density function in Fig. 2.3.

```
x=np.arange(0,20,0.1)
for i in range(1,11):
    plt.plot(x,stats.chi2.pdf(x,i),label='{}'.format(i))
plt.legend(loc='upper_right')
```

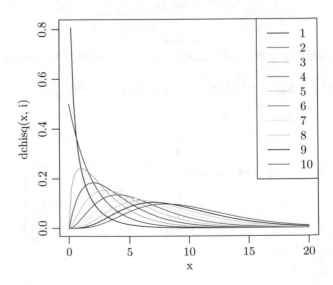

Fig. 2.3 χ^2 distributions with 1 to 10 degrees of freedom

2.5 Hypothesis Testing for $\hat{\beta}_j \neq 0$

In this section, we consider whether each of the $\hat{\beta}_j$, $j = 0, 1, \ldots, p$, is zero or not based on the data.

Without loss of generality, we assume that the values of $x_1, \ldots, x_N \in \mathbb{R}^p$ (row vectors) and $\beta \in \mathbb{R}^{p+1}$ are fixed. However, due to fluctuations in the N random variables $\epsilon_1, \ldots, \epsilon_N$, we may regard that the values

$$y_1 = \beta_0 + x_1[\beta_1, \ldots, \beta_p]^T + \epsilon_1, \ldots, \quad y_N = \beta_0 + x_N[\beta_1, \ldots, \beta_p]^T + \epsilon_N$$

occurred by chance (Fig. 2.4). In fact, if we observe $y_1, \cdots, y_N$ again, since the randomly occurring $\epsilon_1, \ldots, \epsilon_N$ are not the same, the $y_1, \cdots, y_N$ are different from the previous ones.

In the following, although the value of β_j is unknown for each $j = 0, 1, \ldots, p$, we construct a test statistic T that follows a t distribution with $N - p - 1$ degrees of freedom as defined below when we assume $\beta_j = 0$. If the actual value of T is rare under the assumption $\beta_j = 0$, we decide that the hypothesis $\beta_j = 0$ should be rejected.

What we mean by a t distribution with m degrees of freedom is that the distribution of the random variable $T := U/\sqrt{V/m}$ such that $U \sim N(0, 1)$, $V \sim \chi_m^2$ (the χ^2 distribution of degree of freedom m), and U and V are independent.

For each degree of freedom up to m, we depict the graph of the probability density function of the t distribution as in Fig. 2.5. The t distribution is symmetric, its center is at zero, and it approaches the standard Gaussian distribution as the number of degrees of freedom m grows.

Example 22 We allow the degrees of freedom of the t distribution to vary and compare these distributions with the standard Gaussian distribution.

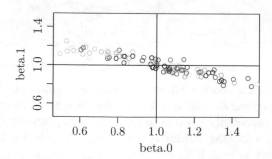

Fig. 2.4 We fix $p = 1$, $N = 100$, and $x_1, \ldots, x_N \sim N(2, 1)$, generate $\epsilon_1, \ldots, \epsilon_N \sim N(0, 1)$, and estimate the intercept β_0 and slope β_1 from $x_1, \ldots, x_N$ and $y_1 = x_1 + 1 + \epsilon_1, \ldots, y_N = x_N + 1 + \epsilon_N$. We repeat the procedure one hundred times and find that the $(\hat{\beta}_0, \hat{\beta}_1)$ values are different

Fig. 2.5 t distributions with 1 to 10 degrees of freedom. The thick line shows the standard Gaussian distribution

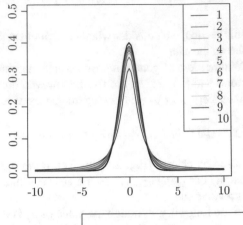

Fig. 2.6 Acceptance and rejection regions for hypothesis testing

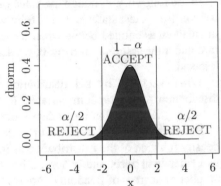

```
x=np.arange(-10,10,0.1)
plt.plot(x,stats.norm.pdf(x,0,1),label="Normal_",c="black",linewidth=1)
for i in range(1,11):
    plt.plot(x,stats.t.pdf(x,i),label='{}'.format(i),linewidth=0.8)
plt.legend(loc='upper_right')
plt.title("changes_of_t_distribution_by_degree_of_freedom")
```

```
Text(0.5, 1.0, 'changes of t distribution by degree of freedom')
```

The hypothesis test constructed here is to set the significance level (e.g., $\alpha = 0.01, 0.05$) and to reject the null hypothesis if the value of T is outside of the range that occurs with probability $1 - \alpha$ as in Fig. 2.6. More precisely, if T is either too large or too small so that the probability is within $\alpha/2$ from both extremes, we reject the null hypothesis $\beta_j = 0$. If $\beta_j = 0$ is true, since $T \sim t_{N-p-1}$, it is rare that T will be far from the center.

We estimate σ in (2.12) and the standard deviation of $\hat{\beta}_j$ by

$$\hat{\sigma} := \sqrt{\frac{RSS}{N - p - 1}}$$

and

$$SE(\hat{\beta}_j) := \hat{\sigma}\sqrt{B_j},$$

respectively, where B_j is the j-th diagonal element of $(X^T X)^{-1}$.

Example 23   For $p = 1$, since

$$X^T X = \begin{bmatrix} 1 & \cdots & 1 \\ x_1 & \cdots & x_N \end{bmatrix} \begin{bmatrix} 1 & x_1 \\ \vdots & \vdots \\ 1 & x_N \end{bmatrix} = N \begin{bmatrix} 1 & \bar{x} \\ \bar{x} & \dfrac{1}{N}\displaystyle\sum_{i=1}^{N} x_i^2 \end{bmatrix},$$

the inverse is

$$(X^T X)^{-1} = \frac{1}{\displaystyle\sum_{i=1}^{N}(x_i - \bar{x})^2} \begin{bmatrix} \dfrac{1}{N}\displaystyle\sum_{i=1}^{N} x_i^2 & -\bar{x} \\ -\bar{x} & 1 \end{bmatrix},$$

which means that

$$B_0 = \frac{\dfrac{1}{N}\displaystyle\sum_{i=1}^{N} x_i^2}{\displaystyle\sum_{i=1}^{N}(x_i - \bar{x})^2} \quad \text{and} \quad B_1 = \frac{1}{\displaystyle\sum_{i=1}^{N}(x_i - \bar{x})^2}.$$

For $B = (X^T X)^{-1}$, $B\sigma^2$ is the covariance matrix of $\hat{\beta}$, and $B_j\sigma^2$ is the variance of $\hat{\beta}_j$. Thus, we may regard $B_j\hat{\sigma}^2$ as an estimate of $B_j\sigma^2$. For $\beta_0 = 1$ and $\beta_1 = 1$, we estimate $\hat{\beta}_0$ and $\hat{\beta}_1$ from $N = 100$ data points. We repeated the process 100 times and plotted them in Fig. 2.4.

```
N=100; p=1
iter_num=100
for i in range(iter_num):
    x=randn(N)+2   # mean=2, var=1
    e=randn(N)
    y=x+1+e
    b_1,b_0=min_sq(x,y)
    plt.scatter(b_0,b_1)
plt.axhline(y=1.0,c="black",linewidth=0.5)
plt.axvline(x=1.0,c="black",linewidth=0.5)
plt.xlabel('beta_0')
plt.ylabel('beta_1')
```

```
Text(0, 0.5, 'beta_1')
```

Because $\bar{x}$ is positive, the correlation between $\hat{\beta}_0$ and $\hat{\beta}_1$ is negative.

In the following, we show

$$t = \frac{\hat{\beta}_j - \beta_j}{SE(\hat{\beta}_j)} \sim t_{N-p-1}. \tag{2.19}$$

To this end, from the definition of the t distribution, we have

$$\frac{\hat{\beta}_j - \beta_j}{SE(\hat{\beta}_j)} = \frac{\hat{\beta}_j - \beta_j}{\sqrt{B_j}\sigma} / \sqrt{\frac{RSS/\sigma^2}{N-p-1}}.$$

Thus, from (2.15)–(2.18), we have

$$U := \frac{\hat{\beta}_j - \beta_j}{\sqrt{B_j}\sigma} \sim N(0,1) \text{ and } V := \frac{RSS}{\sigma^2} \sim \chi^2_{N-p-1} .$$

Hence, it remains to be shown that U and V are independent. In particular, since RSS depends only on $y - \hat{y}$, it is sufficient to show that $y - \hat{y}$ and $\hat{\beta} - \beta$ are independent. To this end, if we note that

$$(\hat{\beta} - \beta)(y - \hat{y})^T = (X^T X)^{-1} X^T \epsilon \epsilon^T (I - H) ,$$

from $E\epsilon\epsilon^T = \sigma^2 I$ and $HX = X$, we have

$$E(\hat{\beta} - \beta)(y - \hat{y})^T = 0 .$$

Since both $y - \hat{y} = (I - H)\epsilon$ and $\hat{\beta} - \beta$ follow Gaussian distributions, zero covariance between them means that they are independent (Proposition 12), which completes the proof.

Example 24 We wish to perform a hypothesis test for a null hypothesis $H_0 : \beta_j = 0$ and its alternative $H_1 : \beta_j \neq 0$. For $p = 1$ and using

$$t = \frac{\hat{\beta}_j - 0}{SE(\hat{\beta}_j)} \sim t_{N-p-1}$$

under H_0, we construct the following procedure in which the function `stats.t.cdf(x,m)` returns $\int_{-\infty}^{x} f_m(t)dt$, where f_m is the probability density function of a t distribution with m degrees of freedom. We compare the output with the output obtained via the `lm` function in the R environment.

```
N=100
x=randn(N); y=randn(N)
beta_1,beta_0=min_sq(x,y)
RSS=np.linalg.norm(y-beta_0-beta_1*x)**2
RSE=np.sqrt(RSS/(N-1-1))
B_0=(x.T@x/N)/np.linalg.norm(x-np.mean(x))**2
B_1=1/np.linalg.norm(x-np.mean(x))**2
se_0=RSE*np.sqrt(B_0)
se_1=RSE*np.sqrt(B_1)
t_0=beta_0/se_0
t_1=beta_1/se_1
p_0=2*(1-stats.t.cdf(np.abs(t_0),N-2))
p_1=2*(1-stats.t.cdf(np.abs(t_1),N-2))
```

```
beta_0,se_0,t_0,p_0 # intercept
```

```
(-0.007650428118828838,
 0.09826142188565655,
 -0.0778579016262494,
 0.9380998328599441)
```

```
beta_1,se_1,t_1,p_1 # coefficient
```

```
(0.03949448841467844,
 0.10414969655462533,
 0.37920886686370736,
 0.7053531714456662)
```

In Python we usually use scikit-learn.

```
from sklearn import linear_model
```

```
reg=linear_model.LinearRegression()
x=x.reshape(-1,1) #  we need to indicate the size of the arrangement in
   sklearn
y=y.reshape(-1,1) #  If we set one of the dimensions and set the other to
   -1, it will automatically adjust itself.
reg.fit(x,y) # execution
```

```
LinearRegression(copy_X=True,fit_intercept=True,n_jobs=None,
        normalize=False)
```

```
reg.coef_,reg.intercept_  # coefficient; beta_1, intercept;  beta_0
```

```
(array([[0.03949449]]), array([-0.00765043]))
```

Now let us use a module called statsmodels to see the details of the results: add all 1's to the left column of X.

```
import statsmodels.api as sm
```

```
X=np.insert(x,0,1,axis=1)
model=sm.OLS(y,X)
res=model.fit()
print(res.summary())
```

```
                            OLS Regression Results
==============================================================================
Dep. Variable:                      y    R-squared:                    0.001
Model:                            OLS    Adj. R-squared:              -0.009
Method:                 Least Squares    F-statistic:                 0.1438
Date:                Wed, 12 Feb 2020    Prob (F-statistic):           0.705
Time:                        14:27:19    Log-Likelihood:             -139.12
No. Observations:                 100    AIC:                          282.2
Df Residuals:                      98    BIC:                          287.5
Df Model:                           1
Covariance Type:            nonrobust
==============================================================================
                 coef     std err          t      P>|t|      [0.025      0.975]
------------------------------------------------------------------------------
const         -0.0077       0.098     -0.078      0.938      -0.203       0.187
x1             0.0395       0.104      0.379      0.705      -0.167       0.246
==============================================================================
Omnibus:                        1.015    Durbin-Watson:                2.182
Prob(Omnibus):                  0.602    Jarque-Bera (JB):             0.534
Skew:                          -0.086    Prob(JB):                     0.766
Kurtosis:                       3.314    Cond. No.                      1.06
==============================================================================
```

Here, we have $RSS = 1.072$ ($df = N - p - 1 = 98$), and the coefficient of determination is 0.02232. For the definition of the adjusted coefficient of determination, see Sect. 2.6.

Example 25 We repeat the estimation $\hat{\beta}_1$ in Example 24 one thousand times ($r = 1000$) to construct the histogram of $\hat{\beta}_1/SE(\beta_1)$. In the following procedure, we compute the quantity beta_1/se_1, and accumulate them as a vector of size r in T. First, we generate the data that follow the null hypothesis $\beta\_1 = 0$ (Fig. 2.7, left).

```
N=100; r=1000
T=[]
for i in range(r):
    x=randn(N); y=randn(N)
    beta_1,beta_0=min_sq(x,y)
    pre_y=beta_0+beta_1*x    # the predicted value of y
    RSS=np.linalg.norm(y-beta_0-beta_1*x)**2
    RSE=np.sqrt(RSS/(N-1-1))
    B_0=(x.T@x/N)/np.linalg.norm(x-np.mean(x))**2
    B_1=1/np.linalg.norm(x-np.mean(x))**2
    se_1=RSE*np.sqrt(B_1)
    T.append(beta_1/se_1)

plt.hist(T,bins=20,range=(-3,3),density=True)
x=np.linspace(-4,4,400)
plt.plot(x,stats.t.pdf(x,98))
plt.title("the_null_hypothesis_holds.")
plt.xlabel('the_value_of_t')
plt.ylabel('probability_density')
```

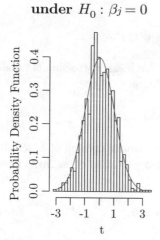
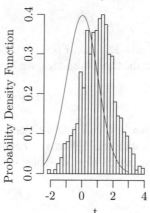

Fig. 2.7 Distribution of $\hat{\beta}_1/SE(\hat{\beta}_1)$ under the null hypothesis $\beta_1 = 0$ (left) and under $\beta_1 = 0.1$ (Right)

```
Text(0, 0.5, 'probability density')
```

Next, we generate data that do not follow the null hypothesis ($\beta_1 = 0.1$) and estimate the model with them, replacing `y=randn(N)` with `y=0.1*x+randn(N)`
(Fig. 2.7, Right).

2.6 Coefficient of Determination and the Detection of Collinearity

In the following, we define a matrix $W \in \mathbb{R}^{N \times N}$ such that all the elements are $1/N$.

Thus, all the elements of $Wy \in \mathbb{R}^N$ are $\bar{y} = \dfrac{1}{N} \sum_{i=1}^{N} y_i$ for $y_1, \ldots, y_N \in \mathbb{R}$.

As we have defined the residual sum of squares as

$$RSS = \|\hat{y} - y\|^2 = \|(I - H)\epsilon\|^2 = \|(I - H)y\|^2 \qquad ,$$

we define the explained sum of squares

$$ESS := \|\hat{y} - \bar{y}\|^2 = \|\hat{y} - Wy\|^2 = \|(H - W)y\|^2$$

and the total sum of squares

$$TSS := \|y - \bar{y}\|^2 = \|(I - W)y\|^2 \quad .$$

If RSS is much less than TSS, we may regard that linear regression is suitable for the data. For the three measures, we have the relation

$$TSS = RSS + ESS. \tag{2.20}$$

Since we have $HX = X$ and the elements in the leftmost column of X are all ones, any all one vector multiplied by a constant is an eigenvector of eigenvalues of one, which means that $HW = W$. Thus, we have

$$(I - H)(H - W) = 0. \tag{2.21}$$

If we square both sides of $(I - W)y = (I - H)y + (H - W)y$, from (2.21), we have $\|(I - W)y\|^2 = \|(I - H)y\|^2 + \|(H - W)y\|^2$. Moreover, we can show that RSS and ESS are independent. To this end, we notice that the covariance matrix between $(I - H)\epsilon$ and $(H - W)y = (H - W)X\beta + (H - W)\epsilon$ is equal to that of $(I - H)\epsilon$ and $(H - W)\epsilon$. In fact, $(H - W)X\beta \in \mathbb{R}^N$ does not fluctuate and is not random. Thus, we may remove it when we compute the covariance matrix. Then, from (2.21), the covariance matrix $E(I - H)\epsilon\epsilon^T(H - W)$ is a zero matrix. Because RSS and ESS follow Gaussian distributions, they are independent (Proposition 12).

We refer to

$$R^2 = \frac{ESS}{TSS} = 1 - \frac{RSS}{TSS}$$

as to the coefficient of determination. As we will see later, for single regression ($p = 1$), the value of R^2 coincides with the square of the sample-based correlation coefficient

$$\hat{\rho} := \frac{\displaystyle\sum_{i=1}^{N}(x_i - \bar{x})(y_i - \bar{y})}{\sqrt{\displaystyle\sum_{i=1}^{N}(x_i - \bar{x})^2 \sum_{i=1}^{N}(y_i - \bar{y})^2}} \quad .$$

In this sense, the coefficient of determination expresses (nonnegative) correlation between the covariates and response. In fact, for $p = 1$, from $\hat{y} = \hat{\beta}_0 + \hat{\beta}_1 x$ and (2.5), we have $\hat{y} - \bar{y} = \hat{\beta}_1(x - \bar{x})$. Hence, from (2.4) and $\|x - \bar{x}\|^2 = \displaystyle\sum_{i=1}^{N}(x_i - \bar{x})^2$

and $\|y - \bar{y}\|^2 = \sum_{i=1}^{N}(y_i - \bar{y})^2$, we have

$$\frac{ESS}{TSS} = \frac{\hat{\beta}_1^2\|x - \bar{x}\|^2}{\|y - \bar{y}\|^2} = \left\{\frac{\sum_{i=1}^{N}(x_i - \bar{x})(y_i - \bar{y})}{\sum_{i=1}^{N}(x_i - \bar{x})^2}\right\}^2 \frac{\sum_{i=1}^{N}(x_i - \bar{x})^2}{\sum_{i=1}^{N}(y_i - \bar{y})^2}$$

$$= \frac{\left\{\sum_{i=1}^{N}(x_i - \bar{x})(y_i - \bar{y})\right\}^2}{\sum_{i=1}^{N}(x_i - \bar{x})^2 \sum_{i=1}^{N}(y_i - \bar{y})^2} = \hat{\rho}^2 .$$

We sometimes use a variant of the coefficient of determination (the adjusted coefficient of determination) such that RSS and TSS are divided by $N - p - 1$ and $N - 1$, respectively:

$$1 - \frac{RSS/(N - p - 1)}{TSS/(N - 1)}. \tag{2.22}$$

If p is large, the adjusted coefficient of determination is smaller than the non-adjusted counterpart. For the regular coefficient of determination, the larger the number of covariates, the better the line fits the data. However, for adjustment covariates, unnecessary covariates that are not removed are penalized.

Example 26 We construct a function to obtain the coefficient of determination and calculate it for actual data.

```
def R2(x,y):
    n=x.shape[0]
    xx=np.insert(x,0,1,axis=1)
    beta=np.linalg.inv(xx.T@xx)@xx.T@y
    y_hat=xx@beta
    y_bar=np.mean(y)
    RSS=np.linalg.norm(y-y_hat)**2
    TSS=np.linalg.norm(y-y_bar)**2
    return 1-RSS/TSS
```

```
N=100; m=2
x=randn(N,m)
y=randn(N)
R2(x,y)
```

```
0.03530233580996256
```

```
# If it is "one" variable, R^2 is the square of the correlation.
x=randn(N,1)
y=randn(N)
R2(x,y)
```

```
0.033782723309598084
```

```
xx=x.reshape(N)
np.corrcoef(xx,y)
```

```
array([[1.         , 0.18380077],
       [0.18380077, 1.         ]])
```

```
np.corrcoef(xx,y)[0,1]**2    # The square of the correlation
```

```
0.033782723309598084
```

While the coefficient of determination expresses how well the covariates explain the response variable, it takes a maximum value of one. We also use VIFs (variance inflation factors), which measures the redundancy of each covariate when the other covariates are present:

$$VIF := \frac{1}{1 - R^2_{X_j|X_{-j}}} ,$$

where $R^2_{X_j|X_{-j}}$ is the coefficient of determination when the j-th variable is the response and the other $p - 1$ variables are covariates in $X \in \mathbb{R}^{N \times p}$ ($y \in \mathbb{R}^N$ is not used when the VIF is computed). The larger the VIF, the better the covariate is explained by the other covariates, which means that the j-th covariate is redundant. The minimum value of VIF is one, and we say that the collinearity of a covariate is strong when its VIF value is large.

Example 27 We installed the Python library `sklearn`, and computed the VIF for the Boston dataset.

```
from sklearn.datasets import load_boston
```

```
boston=load_boston()
x=boston.data
x.shape
```

```
(506, 13)
```

```
def VIF(x):
    p=x.shape[1]
```

```
values=[]
for j in range(p):
    S=list(set(range(p))-{j})
    values.append(1/(1-R2(x[:,S],x[:,j])))
return values
```

```
VIF(x)
```

```
array([1.79219155, 2.29875818, 3.99159642, 1.07399533, 4.39371985,
       1.93374444, 3.10082551, 3.95594491, 7.48449634, 9.00855395,
       1.79908405, 1.34852108, 2.94149108])
```

2.7 Confidence and Prediction Intervals

Thus far, we have showed how to obtain the estimate $\hat{\beta}$ of $\beta \in \mathbb{R}^{p+1}$. In other words, from (2.19), we obtain[2] the confidence interval of $\hat{\beta}$ as follows:

$$\beta_i = \hat{\beta}_i \pm t_{N-p-1}(\alpha/2)SE(\hat{\beta}_i)$$

for $i = 0, 1, \ldots, p$, where $t_{N-p-1}(\alpha/2)$ is the t-statistic such that $\alpha/2 = \int_t^\infty f(u)du$ for the probability density function f.

In this section, we also wish to obtain the confidence interval of $x_*\hat{\beta}$ for another point $x_* \in \mathbb{R}^{p+1}$ (a row vector whose first element is one), which is different from the $x_1, \ldots, x_N$ used for estimation. Then, the average and variance of $x_*\hat{\beta}$ are $E[x_*\hat{\beta}] = x_* E[\hat{\beta}]$ and

$$V[x_*\hat{\beta}] = x_* V(\hat{\beta})x_*^T = \sigma^2 x_*(X^T X)^{-1}x_*^T ,$$

respectively, where σ^2 is the variance of ϵ_i, $i = 1, \ldots, N$. As we derived before, if we define

$$\hat{\sigma} := \sqrt{RSS/(N-p-1)}, \quad SE(x_*\hat{\beta}) := \hat{\sigma}\sqrt{x_*(X^T X)^{-1}x_*^T} ,$$

then we can show that

$$C := \frac{x_*\hat{\beta} - x_*\beta}{SE(x_*\hat{\beta})} = \frac{x_*\hat{\beta} - x_*\beta}{\hat{\sigma}\sqrt{x_*(X^T X)^{-1}x_*^T}}$$

$$= \frac{x_*\hat{\beta} - x_*\beta}{\sigma\sqrt{x_*(X^T X)^{-1}x_*^T}} \bigg/ \sqrt{\frac{RSS}{\sigma^2} \bigg/ (N-p-1)}$$

follows a t distribution with $N - p - 1$ degrees of freedom. In fact, the numerator follows the $N(0, 1)$ distribution, and $\dfrac{RSS}{\sigma^2}$ follows a χ^2_{N-p-1} distribution. More-

---

[2] We write $\hat{\xi} - \gamma \leq \xi \leq \hat{\xi} + \gamma$ as $\xi = \hat{\xi} \pm \gamma$, where $\hat{\xi}$ is an unbiased estimator of ξ.

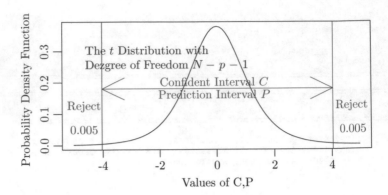

Fig. 2.8 The confidence and prediction intervals are obtained based on the fact that those intervals are the ranges with probability $1 - \alpha$, excluding the tails, where we set $\alpha = 0.01$

over, as we derived before, RSS and $\hat{\beta} - \beta$ are independent. Thus, the proof of $C \sim t_{N-p-1}$ is completed.

On the other hand, if we need to consider the noise ϵ as well as the estimated $x_*\hat{\beta}$ in the evaluation, we consider the variance in the difference between $x_*\hat{\beta}$ and $y_* := x_*\beta + \epsilon$:

$$V[x_*\hat{\beta} - (x_*\beta + \epsilon)] = V[x_*(\hat{\beta} - \beta)] + V[\epsilon] = \sigma^2 x_*(X^T X)^{-1}x_*^T + \sigma^2 .$$

Similarly, we can derive the following:

$$P := \frac{x_*\hat{\beta} - y_*}{SE(x_*\hat{\beta} - y_*)} = \frac{x_*\hat{\beta} - y_*}{\sigma(1 + \sqrt{x_*(X^T X)^{-1}x_*^T})} \bigg/ \sqrt{\frac{RSS}{\sigma^2} \bigg/ (N-p-1)} \sim t_{N-p-1}.$$

Hence, with probability α, we obtain the confidence and prediction intervals, respectively, as follows (Figs. 2.8 and 2.9):

$$x_*\beta = x_*\hat{\beta} \pm t_{N-p-1}(\alpha/2)\hat{\sigma}\sqrt{x_*(X^T X)^{-1}x_*^T}$$

$$y_* = x_*\hat{\beta} \pm t_{N-p-1}(\alpha/2)\hat{\sigma}\sqrt{1 + x_*(X^T X)^{-1}x_*^T}.$$

Example 28 We do not just fit the points to a line via the least squares method; we also draw the confidence interval that surrounds the line and the prediction interval that surrounds both the fitted line and the confidence interval.

```
N=100; p=1
X=randn(N,p)
X=np.insert(X,0,1,axis=1)
beta=np.array([1,1])
epsilon=randn(N)
y=X@beta+epsilon
```

Fig. 2.9 The line obtained via the least squares method. The confidence and prediction intervals are shown by the solid and dashed lines, respectively. In general, the prediction interval lies outside of the confidence interval

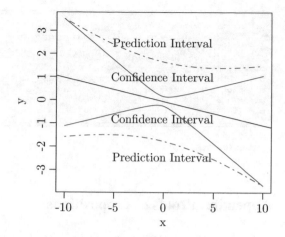

```
# definiton of f(x) and g(x)
U=np.linalg.inv(X.T@X)
beta_hat=U@X.T@y
RSS=np.linalg.norm(y-X@beta_hat)**2
RSE=np.sqrt(RSS/(N-p-1))
alpha=0.05
```

```
def f(x,a):   ## a=0 means confidence , a=1 means prediction
    x=np.array([1,x])
    # stats.t.ppf(0.975,df=N-p-1) # The point at which the cumulative
        probability is 1 - alpha /2
    range=stats.t.ppf(0.975,df=N-p-1)*RSE*np.sqrt(a+x@U@x.T)
    lower=x@beta_hat-range
    upper=x@beta_hat+range
    return ([lower,upper])
```

```
# example
stats.t.ppf(0.975,df=1)   # the point Corresponding to  p
```

```
12.706204736432095
```

```
x_seq=np.arange(-10,10,0.1)
# Confidence interval
lower_seq1=[]; upper_seq1=[]
for i in range(len(x_seq)):
    lower_seq1.append(f(x_seq[i],0)[0]); upper_seq1.append(f(x_seq[i],0)[1])
# prediction interval
lower_seq2=[]; upper_seq2=[]
for i in range(len(x_seq)):
    lower_seq2.append(f(x_seq[i],1)[0]); upper_seq2.append(f(x_seq[i],1)[1])
# Predicted value by regression
yy=beta_hat[0]+beta_hat[1]*x_seq
```

```
plt.xlim(np.min(x_seq),np.max(x_seq))
plt.ylim(np.min(lower_seq1),np.max(upper_seq1))
plt.plot(x_seq,yy,c="black")
plt.plot(x_seq,lower_seq1,c="blue")
plt.plot(x_seq,upper_seq1,c="red")
plt.plot(x_seq,lower_seq2,c="blue",linestyle="dashed")
plt.plot(x_seq,upper_seq2,c="red",linestyle="dashed")
plt.xlabel("x")
plt.ylabel("y")
```

```
Text(0, 0.5, 'y')
```

Appendix: Proofs of Propositions

Proposition 12 *Two Gaussian random variables are independent if and only if their covariance is zero.*

Proof Let $X \sim N(\mu_X, \sigma_X^2)$ and $Y \sim N(\mu_Y, \sigma_Y^2)$, and let $E[\cdot]$ be the expectation operation. If we let

$$\rho := \frac{E(X - \mu_X)(Y - \mu_Y)}{\sqrt{E(X - \mu_X)^2}\sqrt{E(Y - \mu_Y)^2}} \qquad (2.23)$$

and define the independence of X and Y by the property $f_X(x) f_Y(y) = f_{XY}(x, y)$ for all $x, y \in \mathbb{R}$, where

$$f_X(x) = \frac{1}{\sqrt{2\pi}\sigma_X} \exp\left\{ -\frac{1}{2\sigma_X^2}(x - \mu_X)^2 \right\}$$

$$f_Y(y) = \frac{1}{\sqrt{2\pi}\sigma_Y} \exp\left\{ -\frac{1}{2\sigma_Y^2}(y - \mu_Y)^2 \right\}$$

$$f_{XY}(x, y) = \frac{1}{2\pi\sigma_X\sigma_Y\sqrt{1 - \rho^2}}$$

$$\times \exp\left\{ -\frac{1}{2(1 - \rho^2)} \left[\left(\frac{x - \mu_X}{\sigma_X}\right)^2 - 2\rho\left(\frac{x - \mu_X}{\sigma_X}\right)\left(\frac{y - \mu_Y}{\sigma_Y}\right) \right.\right.$$

$$\left.\left. + \left(\frac{x - \mu_X}{\sigma_X}\right)^2 \right] \right\} ,$$

then $\rho = 0 \implies f_{XY}(x, y) = f_X(x)f_Y(y)$. On the other hand, if $f_{XY}(x, y) = f_X(x)f_Y(y)$, then we can write the numerator of ρ in (2.23) as follows:

$$\int_{-\infty}^{\infty}\int_{-\infty}^{\infty} (x - \mu_X)(y - \mu_Y)f_{XY}(x, y)dxdy$$
$$= \int_{-\infty}^{\infty}(x - \mu_X)f_X(x)dx \int_{-\infty}^{\infty}(y - \mu_Y)f_Y(y)dy$$
$$= 0,$$

which means that $\rho = 0 \impliedby f_{XY}(x, y) = f_X(x)f_Y(y)$. □

Proposition 13 The eigenvalues of H and $I - H$ are only zeros and ones, and the dimensions of the eigenspaces of H and $I - H$ with eigenvalues one and zero, respectively, are both $p+1$, while the dimensions of the eigenspaces of H and $I - H$ with eigenvalues of zero and one, respectively, are both $N - p - 1$.

Proof Using Proposition 4, from $H = X(X^TX)^{-1}X^T$ and rank$(X) = p + 1$, we have

$$\text{rank}(H) \leq \min\{\text{rank}(X(X^TX)^{-1}), \text{rank}(X)\} \leq \text{rank}(X) = p + 1.$$

On the other hand, from Proposition 4 and $HX = X$, rank$(X) = p + 1$, we have

$$\text{rank}(H) \geq \text{rank}(HX) = \text{rank}(X) = p + 1.$$

Therefore, we have rank$(H) = p + 1$. Moreover, from $HX = X$, the columns of X are the basis of the image of H and the eigenvectors of H for an eigenvalue of one. Since the dimension of the image of H is $p + 1$, the dimension of the kernel is $N - p - 1$ (the eigenspace of an eigenvalue of zero). Moreover, for an arbitrary $x \in \mathbb{R}^{p+1}$, we have $(I - H)x = 0 \iff Hx = x$ and $(I - H)x = x \iff Hx = 0$, which means that the eigenspaces of H and $I - H$ for eigenvalues of zero and one are the same as the eigenspaces of $I - H$ and H for eigenvalues one and zero, respectively. □

Exercises 1–18

1. For a given $x_1, \ldots, x_N, y_1, \ldots, y_N \in \mathbb{R}$, let $\hat{\beta}_0, \hat{\beta}_1$ be the $\beta_0, \beta_1 \in \mathbb{R}$ that minimizes $L := \sum_{i=1}^{N}(y_i - \beta_0 - \beta_1 x_i)^2$. Show the following equations, where $\bar{x}$ and $\bar{y}$ are defined by $\frac{1}{N}\sum_{i=1}^{N}x_i$ and $\frac{1}{N}\sum_{i=1}^{N}y_i$.

(a) $\hat{\beta}_0 + \hat{\beta}_1 \bar{x} = \bar{y}$

(b) Unless $x_1 = \ldots = x_N$,

$$\hat{\beta}_1 = \frac{\displaystyle\sum_{i=1}^{N}(x_i - \bar{x})(y_i - \bar{y})}{\displaystyle\sum_{i=1}^{N}(x_i - \bar{x})^2}$$

Hint: Item (a) is obtained from $\dfrac{\partial L}{\partial \beta_0} = 0$. For (b), substitute (a) into $\dfrac{\partial L}{\partial \beta_1} =$

$$-2\sum_{i=1}^{N} x_i(y_i - \beta_0 - \beta_1 x_i) = 0$$ and eliminate β_0. Then, solve it w.r.t. β_1 first and

obtain β_0 later.

2. We consider the line l with the intercept $\hat{\beta}_0$ and slope $\hat{\beta}_1$ obtained in Problem 1. Find the intercept and slope of the shifted line l' from the data $x_1 - \bar{x}, \ldots, x_N - \bar{x}$ and $y_1 - \bar{y}, \ldots, y_N - \bar{y}$. How do we obtain the intercept and slope of l from those of the shifted line l'?

3. We wish to visualize the relation between the lines l, l' in Problem 2. Fill Blanks (1) and (2) below and draw the graph.

```python
def min_sq(x,y):   # function for finding the intercept and coefficient of
    the least-squares
    x_bar,y_bar=np.mean(x),np.mean(y)
    beta_1=np.dot(x-x_bar,y-y_bar)/np.linalg.norm(x-x_bar)**2
    beta_0=y_bar-beta_1*x_bar
    return [beta_1,beta_0]
```

```python
N=100
a=np.random.normal(loc=2,scale=1)   # mean, variance, size
b=randn(1) # coefficient
x=randn(N)
y=a*x+b+randn(N)

a1,b1=min_sq(x,y)              # estimating
xx=x-# blank(1) #
yy=y-# blank(2) #
a2,b2=min_sq(xx,yy)            # estimating after centering
```

```
(1.7865393047324676, 1.067565008452225e-16)
```

```python
x_seq=np.arange(-5,5,0.1)
y_pre=x_seq*a1+b1
yy_pre=x_seq*a2+b2
plt.scatter(x,y,c="black")
plt.axhline(y=0,c="black",linewidth=0.5)
plt.axvline(x=0,c="black",linewidth=0.5)
plt.plot(x_seq,y_pre,c="blue",label="before_centering")
plt.plot(x_seq,yy_pre,c="orange",label="after_centering")
plt.legend(loc="upper_left")
```

4. Let m, n be positive integers. Suppose that the matrix $A \in \mathbb{R}^{m \times m}$ can be written by $A = B^T B$ for some $B \in \mathbb{R}^{n \times m}$.

   (a) Show that $Az = 0 \iff Bz = 0$ for arbitrary $z \in \mathbb{R}^m$. Hint: Use $Az = 0 \implies z^T B^T Bz = 0 \implies \|Bz\|^2 = 0$.
   (b) Show that the ranks of A and B are equal. Hint: Because the kernels of A and B are equal so are the dimensions (ranks) of the images.

In the following, the leftmost column of $X \in \mathbb{R}^{N \times (p+1)}$ consists of all ones.

5. For each of the following cases, show that $X^T X$ is not invertible:

   (a) $N < p + 1$.
   (b) $N \geq p + 1$ and different columns are equal in X.

In the following, the rank of $X \in \mathbb{R}^{N \times (p+1)}$ is $p + 1$.

6. We wish to obtain $\beta \in \mathbb{R}^{p+1}$ that minimizes $L := \|y - X\beta\|^2$ from $X \in \mathbb{R}^{N \times (p+1)}$, $y \in \mathbb{R}^N$, where $\| \cdot \|$ denotes $\sqrt{\sum_{i=1}^{N} z_i^2}$ for $z = [z_1, \cdots, z_N]^T$.

   (a) Let $x_{i,j}$ be the (i, j)-th element of X. Show that the partial derivative of $L =$
   $$\frac{1}{2} \sum_{i=1}^{N} \left(y_i - \sum_{j=0}^{p} x_{i,j} \beta_j \right)^2$$
   w.r.t. β_j is the j-th element of $-X^T y + X^T X\beta$.

   Hint: The j-th element of $X^T y$ is $\sum_{i=1}^{N} x_{i,j} y_i$, the (j, k)-th element of $X^T X$ is $\sum_{i=1}^{N} x_{i,j} x_{i,k}$, and the j-th element of $X^T X\beta$ is $\sum_{k=0}^{p} \sum_{i=1}^{N} x_{i,j} x_{i,k} \beta_k$.

   (b) Find $\beta \in \mathbb{R}^{p+1}$ such that $\dfrac{\partial L}{\partial \beta} = 0$. In the sequel, we write the value by $\hat{\beta}$.

7. Suppose that the random variable $\hat{\beta}$ is obtained via the procedure in Problem 6, where we assume that $X \in \mathbb{R}^{N \times (p+1)}$ is given and $y \in \mathbb{R}^N$ is generated by $X\beta + \epsilon$ with unknown constants $\beta \in \mathbb{R}^{p+1}$ and $\sigma^2 > 0$ and random variable $\epsilon \sim N(0, \sigma^2 I)$.

   (a) Show $\hat{\beta} = \beta + (X^T X)^{-1} X^T \epsilon$.
   (b) Show that the average of $\hat{\beta}$ coincides with β, i.e., $\hat{\beta}$ is an unbiased estimator.
   (c) Show that the covariance matrix of $\hat{\beta}$ is $E(\hat{\beta} - \beta)(\hat{\beta} - \beta)^T = \sigma^2 (X^T X)^{-1}$.

8. Let $H := X(X^T X)^{-1} X^T \in \mathbb{R}^{N \times N}$ and $\hat{y} := X\hat{\beta}$. Show the following equations:

   (a) $H^2 = H$,
   (b) $(I - H)^2 = I - H$,
   (c) $HX = X$,

(d) $\hat{y} = Hy$,

(e) $y - \hat{y} = (I - H)\epsilon$,

(f) $\|y - \hat{y}\|^2 = \epsilon^T (I - H)\epsilon$.

9. Prove the following statements:

   (a) The dimension of the image, rank, of H is $p + 1$. Hint: We assume that the rank of X is $p + 1$.

   (b) H has eigenspaces of eigenvalues of zero and one, and their dimensions are $N - p - 1$ and $p + 1$, respectively. Hint: The number of columns N in H is the sum of the dimensions of the image and kernel.

   (c) $I - H$ has eigenspaces of eigenvalues of zero and one, and their dimensions are $p + 1$ and $N - p - 1$, respectively. Hint: For an arbitrary $x \in \mathbb{R}^{p+1}$, we have $(I - H)x = 0 \iff Hx = x$ and $(I - H)x = x \iff Hx = 0$.

10. Using the fact that $P(I - H)P^T$ becomes a diagonal matrix such that the first $N - p - 1$ and last $p + 1$ diagonal elements are ones and zeros, respectively, for an orthogonal P, show the following:

   (a) $RSS := \epsilon^T (I - H)\epsilon = \sum_{i=1}^{N-p-1} v_i^2$, where $v := P\epsilon$. Hint: Because P is orthogonal, we have $P^T P = I$. Substitute $\epsilon = P^{-1}v = P^T v$ into the definition of RSS and find that the diagonal elements of $P^T (I - H)P$ are the N eigenvalues. In particular, $I - H$ has $N - p - 1$ and $p + 1$ eigenvalues of zero and one, respectively.

   (b) $Evv^T = \sigma^2 \tilde{I}$. Hint: Use $Evv^T = P(E\epsilon\epsilon^T)P^T$.

   (c) $RSS/\sigma^2 \sim \chi^2_{N-p-1}$

   (the χ^2 distribution with $N - p - 1$ degrees of freedom). Hint: Find the statistical properties from (a) and (b).

   Use the fact that the independence of Gaussian random variables is equivalent to the covariance matrix of them being diagonal, without proving it.

11. (a) Show that $E(\hat{\beta} - \beta)(y - \hat{y})^T = 0$. Hint: Use $(\hat{\beta} - \beta)(y - \hat{y})^T = (X^T X)^{-1} X^T \epsilon\epsilon^T (I - H)$ and $E\epsilon\epsilon^T = \sigma^2 I$.

   (b) Let $B_0, \ldots, B_p$ be the diagonal elements of $(X^T X)^{-1}$. Show that $(\hat{\beta}_i - \beta_i)/(\sqrt{B_i}\sigma)$ and RSS/σ^2 are independent for $i = 0, 1, \ldots, p$. Hint: Since RSS is a function of $y - \hat{y}$, the problem reduces to independence between $y - \hat{y}$ and $\hat{\beta} - \beta$. Because they are Gaussian, it is sufficient to show that the covariance is zero.

   (c) Let $\hat{\sigma} := \sqrt{\dfrac{RSS}{N - p - 1}}$ (the residual standard error, an estimate of σ), and

   $SE(\hat{\beta}_i) := \hat{\sigma}\sqrt{B_i}$ (an estimate of the standard error of $\hat{\beta}_i$). Show that

   $$\frac{\hat{\beta}_i - \beta_i}{SE(\hat{\beta}_i)} \sim t_{N-p-1}, \quad i = 0, 1, \ldots, p$$

   (the t distribution with $N - p - 1$ degrees of freedom). Hint: Derive

$$\frac{\hat{\beta}_i - \beta_i}{SE(\hat{\beta}_i)} = \frac{\hat{\beta}_i - \beta_i}{\sigma\sqrt{B_i}} \bigg/ \sqrt{\frac{RSS}{\sigma^2} \bigg/ (N - p - 1)}$$

and show that the right-hand side follows a t distribution.

(d) When $p = 1$, find B_0 and B_1, letting $(x_{1,1}, \ldots, x_{N,1}) = (x_1, \ldots, x_N)$.
Hint: Derive

$$(X^T X)^{-1} = \frac{1}{\displaystyle\sum_{i=1}^{N}(x_i - \bar{x})^2} \begin{bmatrix} \dfrac{1}{N}\displaystyle\sum_{i=1}^{N} x_i^2 & -\bar{x} \\ -\bar{x} & 1 \end{bmatrix}.$$

Use the fact that independence of Gaussian random variables $U_1, \ldots, U_m$, $V_1, \ldots, V_N$ is equivalent to a covariance matrix of size $m \times n$ being a diagonal matrix, without proving it.

12. We wish to test the null hypothesis $H_0 : \beta_i = 0$ versus its alternative $H_1 : \beta_i \neq 0$. For $p = 1$, we construct the following procedure using the fact that under H_0,

$$t = \frac{\hat{\beta}_i - 0}{SE(\hat{\beta}_i)} \sim t_{N-p-1} ,$$

where the function stats.t.cdf(x,m) returns the value of $\displaystyle\int_x^\infty f_m(t)dt$, where f_m is the probability density function of a t distribution with m degrees of freedom.

```
N=100
x=randn(N); y=randn(N)
beta_1,beta_0=min_sq(x,y)
RSS=np.linalg.norm(y-beta_0-beta_1*x)**2
RSE=np.sqrt(RSS/(N-1-1))
B_0=(x.T@x/N)/np.linalg.norm(x-np.mean(x))**2
B_1=1/np.linalg.norm(x-np.mean(x))**2
se_0=RSE*np.sqrt(B_0)
se_1=RSE*np.sqrt(B_1)
t_0=beta_0/se_0
t_1=beta_1/se_1
p_0=2*(1-stats.t.cdf(np.abs(t_0),N-2))
p_1=2*(1-stats.t.cdf(np.abs(t_1),N-2))
beta_0,se_0 ,t_0 ,p_0  #intercept
beta_1,se_1,t_1,p_1  # coefficient
```

Examine the outputs using the stats_model package and linear_model function in the Python language.

```
from sklearn import linear_model
```

```
reg=linear_model.LinearRegression()
x=x.reshape(-1,1)  #  we need to indicate the size of the arrangement in
    sklearn
y=y.reshape(-1,1)  #  If we set one of the dimensions and set the other to
    -1, it will automatically adjust itself.
reg.fit(x,y) # execute
reg.coef_,reg.intercept_  # coefficient; beta_1, intercept; beta_0
```

```
import statsmodels.api as sm
```

```
X=np.insert(x,0,1,axis=1)
model=sm.OLS(y,X)
res=model.fit()
print(res.summary())
```

13. The following procedure repeats estimating $\hat{\beta}_1$ one thousand times ($r = 1000$) and draws a histogram of $\hat{\beta}_1/SE(\beta_1)$, where `beta_1/se_1` is computed each time from the data, and they are accumulated in the vector T of size r.

```
N=100; r=1000
T=[]
for i in range(r):
    x=randn(N); y=randn(N)
    beta_1,beta_0=min_sq(x,y)
    pre_y=beta_0+beta_1*x # the predicted value of y
    RSS=np.linalg.norm(y-beta_0-beta_1*x)**2
    RSE=np.sqrt(RSS/(N-1-1))
    B_0=(x.T@x/N)/np.linalg.norm(x-np.mean(x))**2
    B_1=1/np.linalg.norm(x-np.mean(x))**2
    se_1=RSE*np.sqrt(B_1)
    T.append(beta_1/se_1)
plt.hist(T,bins=20,range=(-3,3),density=True)
x=np.linspace(-4,4,400)
plt.plot(x,stats.t.pdf(x,1))
plt.title("the_null_hypothesis_holds")
plt.xlabel('the_value_of_t')
plt.ylabel('probability_density')
```

   Replace `y=randn(N)` with `y=0.1*x+randn(N)` and execute it. Furthermore, explain the difference between the two graphs.

14. Suppose that each element of $W \in \mathbb{R}^{N \times N}$ is $1/N$, thus $\bar{y} = \dfrac{1}{N} \displaystyle\sum_{i=1}^{N} y_i = Wy$

   for $y = [y_1, \cdots, y_N]^T$.

   (a) Show that $HW = W$ and $(I - H)(H - W) = 0$. Hint: Because each column of W is an eigenvector of eigenvalue one in H, we have $HW = W$.
   (b) Show that $ESS := \|\hat{y} - \bar{y}\|^2 = \|(H - W)y\|^2$ and $TSS := \|y - \bar{y}\|^2 = \|(I - W)y\|^2$.
   (c) Show that $RSS = \|(I - H)\epsilon\|^2 = \|(I - H)y\|^2$ and ESS are independent Hint: The covariance matrix of $(I - H)\epsilon$ and $(H - W)y$ is that of $(I - H)\epsilon$

and $(H - W)\epsilon$. Evaluate the covariance matrix $E(I - H)\epsilon\epsilon^T(H - W)$. Then, use (a).

(d) Show that $\|(I - W)y\|^2 = \|(I - H)y\|^2 + \|(H - W)y\|^2$, i.e., $TSS = RSS + ESS$. Hint: $(I - W)y = (I - H)y + (H - W)y$.

In the following, we assume that $X \in \mathbb{R}^{N \times p}$ does not contain a vector of size N of all ones in the leftmost column.

15. Given $X \in \mathbb{R}^{N \times p}$ and $y \in \mathbb{R}^N$, we refer to

$$R^2 = \frac{ESS}{TSS} = 1 - \frac{RSS}{TSS}$$

as to the coefficient of determination. For $p = 1$, suppose that we are given $x = [x_1, \ldots, x_N]^T$.

(a) Show that $\hat{y} - \bar{y} = \hat{\beta}_1(x - \bar{x})$. Hint: Use $\hat{y}_i = \hat{\beta}_0 + \hat{\beta}_1 x_i$ and Problem 1(a).

(b) Show that $R^2 = \dfrac{\hat{\beta}_1^2 \|x - \bar{x}\|^2}{\|y - \bar{y}\|^2}$.

(c) For $p = 1$, show that the value of R^2 coincides with the square of the correlation coefficient. Hint: Use $\|x - \bar{x}\|^2 = \sum_{i=1}^{N}(x_i - \bar{x})^2$ and Problem 1(b).

(d) The following function computes the coefficient of determination:

```
def R2(x,y):
    n=x.shape[0]
    xx=np.insert(x,0,1,axis=1)
    beta=np.linalg.inv(xx.T@xx)@xx.T@y
    y_hat=xx@beta
    y_bar=np.mean(y)
    RSS=np.linalg.norm(y-y_hat)**2
    TSS=np.linalg.norm(y-y_bar)**2
    return 1-RSS/TSS
N=100; m=2; x=randn(N,m); y=randn(N); R2(x,y)
```

Let N=100 and m=1, and execute x=randn(N); y=randn(N); R2(x,y); np.corrcoef(x,y)^2.

16. The coefficient of determination expresses how well the covariates explain the response variable, and its maximum value is one. When we evaluate how redundant a covariate is when the other covariates are present, we often use VIFs (variance inflation factors)

$$VIF := \frac{1}{1 - R^2_{X_j|X_{-j}}},$$

where $R^2_{X_j|X_{-j}}$ is the coefficient of determination of the j-th covariate in $X \in \mathbb{R}^{N \times p}$ given the other $p - 1$ covariates ($y \in \mathbb{R}^N$ is not used). The larger the VIF value, the better the covariate is explained by the other covariates (the minimum value is one), which means that the collinearity is strong. Install the sklearn.datasets and compute the VIF values for each variable in the Boston dataset by filling the blank. (Simply execute the following).

```
from sklearn.datasets import load_boston
```

```
boston=load_boston()
p=x.shape[1]; values=[]
for j in range(p):
    S=list(set(range(p))-{j})
    values.append(# blank #)
values
```

17. We can compute the prediction value $x_*\hat{\beta}$ for each $x_* \in \mathbb{R}^{p+1}$ (the row vector whose first value is one), using the estimate $\hat{\beta}$.

   (a) Show that the variance of $x_*\hat{\beta}$ is $\sigma^2 x_*(X^T X)^{-1}x_*^T$. Hint: Use $V(\hat{\beta}) = \sigma^2(X^T X)^{-1}$.

   (b) If we let $SE(x_*^T \hat{\beta}) := \hat{\sigma}\sqrt{x_*(X^T X)^{-1}x_*^T}$, show that

$$\frac{x_*\hat{\beta} - x_*\beta}{SE(x_*\hat{\beta})} \sim t_{N-p-1} \, ,$$

   where $\hat{\sigma} = \sqrt{RSS/(N - p - 1)}$.

   (c) The actual value of y can be expressed by $y_* := x_*\beta + \epsilon$. Thus, the variance of $y_* - x_*\hat{\beta}$ is σ^2 larger. Show that

$$\frac{x_*\hat{\beta} - y_*}{\hat{\sigma}\sqrt{1 + x_*(X^T X)^{-1}x_*^T}} \sim t_{N-p-1} \, .$$

18. From Problem 17, we have

$$x_*^T \hat{\beta} \pm t_{N-p-1}(\alpha/2)\hat{\sigma}\sqrt{x_*^T (X^T X)^{-1}x_*}$$

$$y_* \pm t_{N-p-1}(\alpha/2)\hat{\sigma}\sqrt{1 + x_*^T (X^T X)^{-1}x_*}$$

(the confidence and prediction intervals, respectively), where f is the t distribution with $N - p - 1$ degrees of freedom. $t_{N-p-1}(\alpha/2)$ is the t-statistic such that $\alpha/2 = \int_t^\infty f(u)du$. Suppose that $p = 1$. We wish to draw the confidence and prediction intervals in red and blue, respectively, for $x_* \in \mathbb{R}$. For the confidence interval, we expressed the upper and lower limits by red and blue solid lines, respectively, executing the procedure below. For the prediction interval, define the function $g(x)$ and overlay the upper and lower dotted lines in red and blue on the same graph.

```
N=100; p=1
X=randn(N,p)
X=np.insert(X,0,1,axis=1)
beta=np.array([1,1])
epsilon=randn(N)
y=X@beta+epsilon

# definition of f(x) and g(x)
U=np.linalg.inv(X.T@X)
beta_hat=U@X.T@y
RSS=(y-X@beta_hat).T@(y-X@beta_hat)
RSE=np.sqrt(RSS/(N-p-1))
alpha=0.05
def f(x):
    x=np.array([1,x])
    # stats.t.ppf(0.975,df=N-p-1) # the point at which the cumulative
      probability is 1-alpha/2
    range=stats.t.ppf(0.975,df=N-p-1)*RSE*np.sqrt(x@U@x.T)
    lower=x@beta_hat-range
    upper=x@beta_hat+range
    return ([lower,upper])

x_seq=np.arange(-10,10,0.1)
lower_seq1=[]; upper_seq1=[]
for i in range(len(x_seq)):
    lower_seq1.append(f(x_seq[i],0)[0])
    upper_seq1.append(f(x_seq[i],0)[1])
yy=beta_hat[0]+beta_hat[1]*x_seq

plt.xlim(np.min(x_seq),np.max(x_seq))
plt.ylim(np.min(lower_seq1),np.max(upper_seq1))
plt.plot(x_seq,yy,c="black")
plt.plot(x_seq,lower_seq1,c="blue")
plt.plot(x_seq,upper_seq1,c="red")
plt.plot(x_seq,lower_seq2,c="blue",linestyle="dashed")
plt.plot(x_seq,upper_seq2,c="red",linestyle="dashed")
plt.xlabel("x")
plt.ylabel("y")
```

Chapter 3
Classification

Abstract In this chapter, we consider constructing a classification rule from covariates to a response that takes values from a finite set such as ± 1, figures $0, 1, \cdots, 9$. For example, we wish to classify a postal code from handwritten characters and to make a rule between them. First, we consider logistic regression to minimize the error rate in the test data after constructing a classifier based on the training data. The second approach is to draw borders that separate the regions of the responses with linear and quadratic discriminators and the k-nearest neighbor algorithm. The linear and quadratic discriminations draw linear and quadratic borders, respectively, and both introduce the notion of prior probability to minimize the average error probability. The k-nearest neighbor method searches the border more flexibly than the linear and quadratic discriminators. On the other hand, we take into account the balance of two risks, such as classifying a sick person as healthy and classifying a healthy person as unhealthy. In particular, we consider an alternative approach beyond minimizing the average error probability. The regression method in the previous chapter and the classification method in this chapter are two significant issues in the field of machine learning.

3.1 Logistic Regression

We wish to determine a decision rule from p covariates to a response that takes two values. More precisely, we derive the map $x \in \mathbb{R}^p \to y \in \{-1, 1\}$ from the data $(x_1, y_1), \ldots, (x_N, y_N) \in \mathbb{R}^p \times \{-1, 1\}$ that minimizes the error probability.

© The Author(s), under exclusive license to Springer Nature Singapore Pte Ltd. 2021
J. Suzuki, *Statistical Learning with Math and Python*,
https://doi.org/10.1007/978-981-15-7877-9_3

Fig. 3.1 As the value of β increases, the probability of $y = 1$ increases monotonically and changes greatly from approximately 0 to approximately 1 near $x = 0$

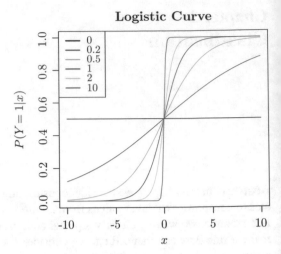

In this section, we assume that for $x \in \mathbb{R}^p$ (row vector), the probabilities of $y = 1$ and $y = -1$ are expressed by[1] $\dfrac{e^{\beta_0 + x\beta}}{1 + e^{\beta_0 + x\beta}}$ and $\dfrac{1}{1 + e^{\beta_0 + x\beta}}$, respectively, for some $\beta_0 \in \mathbb{R}$ and $\beta \in \mathbb{R}^p$ and write the probability of $y \in \{-1, 1\}$ as

$$\frac{1}{1 + e^{-y(\beta_0 + x\beta)}}$$

(logistic regression). To roughly explain the function (the sigmoid function), we draw the graph for $p = 1$, $\beta_0 = 0$, $\beta > 0$, and $y = 1$:

$$f(x) = \frac{1}{1 + e^{-(\beta_0 + x\beta)}} , \ x \in \mathbb{R} .$$

Example 29 We ran the following program, and the graph is shown in Fig. 3.1:

```
def f(x):
    return np.exp(beta_0+beta*x)/(1+np.exp(beta_0+beta*x))
```

```
beta_0=0
beta_seq=np.array([0,0.2,0.5,1,2,10])
x_seq=np.arange(-10,10,0.1)
plt.xlabel("x")
plt.ylabel("P(Y=1|x)")
plt.title("logistic_curve")
for i in range(beta_seq.shape[0]):
    beta=beta_seq[i]
    p=f(x_seq)
    plt.plot(x_seq,p,label='{}'.format(beta))
plt.legend(loc='upper_left')
```

---

[1] In this chapter, instead of $\beta \in \mathbb{R}^{p+1}$, we separate the slope $\beta \in \mathbb{R}^p$ and the intercept $\beta_0 \in \mathbb{R}$.

From

$$f'(x) = \beta \frac{e^{-(\beta_0 + x\beta)}}{(1 + e^{-(\beta_0 + x\beta)})^2} \geq 0$$

$$f''(x) = -\beta^2 \frac{e^{-(\beta_0 + x\beta)}[1 - e^{-(\beta_0 + x\beta)}]}{(1 + e^{-(\beta_0 + x\beta)})^3},$$

we see that $f(x)$ is increasing monotonically and is convex and concave when $x < -\beta_0/\beta$ and $x > -\beta_0/\beta$, respectively; they change at x=0, when $\beta_0 = 0$.

In the following, from the observations $(x_1, y_1), \ldots, (x_N, y_N) \in \mathbb{R}^p \times \{-1, 1\}$, by maximizing the likelihood $\prod_{i=1}^{N} \frac{1}{1 + e^{-y_i(\beta_0 + x_i\beta)}}$ (maximum likelihood), or minimizing the negative log-likelihood:

$$l(\beta_0, \beta) = \sum_{i=1}^{N} \log(1 + v_i), \quad v_i = e^{-y_i(\beta_0 + x_i\beta)}, \quad i = 1, \ldots, N,$$

we obtain the estimate $\beta_0 \in \mathbb{R}$, $\beta \in \mathbb{R}^p$.

Example 30 If the observations are

i	1	2	3	$\cdots$	25
x_i	71.2	29.3	42.3	$\cdots$	25.8
y_i	-1	-1	1	$\cdots$	1

($p = 1, N = 25$), the likelihood to be maximized is

$$\frac{1}{1 + \exp(\beta_0 + 71.2\beta_1)} \cdot \frac{1}{1 + \exp(\beta_0 + 29.3\beta_1)} \cdot \frac{1}{1 + \exp(-\beta_0 - 42.3\beta_1)}$$

$$\cdots \frac{1}{1 + \exp(-\beta_0 - 25.8\beta_1)} \cdot$$

Note that the observations are known, and we determine β_0, β_1 so that the likelihood is maximized.

However, for logistic regression, unlike for linear regression, no formula to obtain the estimates of the coefficients exists.

3.2 Newton–Raphson Method

When we solve the equations such as the partial derivatives of $l(\beta_0, \beta)$ being zero, the Newton–Raphson method is often used. To understand the essence, briefly, we consider the purest example of the use of the Newton–Raphson method. Suppose that we solve $f(x) = 0$ with

$$f(x) = x^2 - 1 .$$

We set an initial value $x = x_0$ and draw the tangent that goes through the point $(x_0, f(x_0))$. If the tangent crosses the x-axis ($y = 0$) at $x = x_1$, then we again draw the tangent that intersects the point $(x_1, f(x_1))$. If we repeat the process, the sequence $x_0, x_1, x_2, \ldots$ approaches the solution of $f(x) = 0$. In general, because the tangent line is $y - f(x_i) = f'(x_i)(x - x_i)$, the intersection with $y = 0$ is

$$x_{i+1} := x_i - \frac{f(x_i)}{f'(x_i)} \tag{3.1}$$

for $i = 0, 1, 2, \cdots$. If more than one solution exists, the solution obtained by the convergence may depend on the initial value of x_0. In the current case, if we set $x_0 = -2$, the solution converges to $x = -1$. In addition, we need to decide when the cycle should be terminated based on some conditions, such as the size of $|x_{i+1} - x_i|$ and the number of repetitions.

Example 31 For $x_0 = 4$, we run the following Python program to obtain the graph in Fig. 3.2. The program repeats the cycle ten times.

```
def f(x):
    return x**2-1
def df(x):
    return 2*x
```

```
x_seq=np.arange(-1,5,0.1)
f_x=f(x_seq)
plt.plot(x_seq,f_x)
plt.axhline(y=0,c="black",linewidth=0.5)
plt.xlabel("x")
plt.ylabel("f(x)")
x=4
for i in range(10):
    X=x; Y=f(x)    # X,Y  before updating
    x=x-f(x)/df(x) # x   after updating
    y=f(x)         # y   after updating
    plt.plot([X,x],[Y,0],c="black",linewidth=0.8)
    plt.plot([X,X],[Y,0],c="black",linestyle="dashed",linewidth=0.8)
    plt.scatter(x,0,c="red")
```

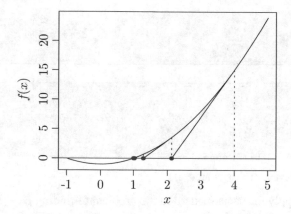

Fig. 3.2 The Newton–Raphson method: starting from $x_0 = 4$, the tangent that goes through $(x_0, f(x_0))$ and crosses the x-axis at x_1, and the tangent that goes through $(x_1, f(x_1))$ and crosses the x-axis at x_2, and so on. The sequence is obtained by the recursion $x_1 = x_0 - f(x_0)/f'(x_0)$, $x_2 = x_1 - f(x_1)/f'(x_1)$, …. The points in the sequence are marked in red

The Newton–Raphson method can even be applied to two variables and two equations: for

$$\begin{cases} f(x, y) = 0 \\ g(x, y) = 0 \end{cases},$$

we can see that (3.1) is extended to

$$\begin{bmatrix} x \\ y \end{bmatrix} \leftarrow \begin{bmatrix} x \\ y \end{bmatrix} - \begin{bmatrix} \dfrac{\partial f(x, y)}{\partial x} & \dfrac{\partial f(x, y)}{\partial y} \\ \dfrac{\partial g(x, y)}{\partial x} & \dfrac{\partial g(x, y)}{\partial y} \end{bmatrix}^{-1} \begin{bmatrix} f(x, y) \\ g(x, y) \end{bmatrix}, \tag{3.2}$$

where the matrix $\begin{bmatrix} \dfrac{\partial f(x, y)}{\partial x} & \dfrac{\partial f(x, y)}{\partial y} \\ \dfrac{\partial g(x, y)}{\partial x} & \dfrac{\partial g(x, y)}{\partial y} \end{bmatrix}$ is called a Jacobian matrix.

Example 32 For $f(x, y) = x^2 + y^2 - 1$ and $g(x, y) = x + y$, if we start searching the solution from $(x, y) = (3, 4)$, the execution is as follows:

```
def f(z):
    return z[0]**2+z[1]**2-1
def dfx(z):
    return 2*z[0]
def dfy(z):
    return 2*z[1]
def g(z):
    return z[0]+z[1]
```

```
def dgx(z):
    return 1
def dgy(z):
    return 1
```

```
z=np.array([3,4]) # initial value
for i in range(10):
    J=np.array([[dfx(z),dfy(z)],[dgx(z),dgy(z)]])
    z=z-np.linalg.inv(J)@np.array([f(z),g(z)])
z
```

array([-0.70710678, 0.70710678])

Then, we apply the same method to the problem of finding $\beta_0 \in \mathbb{R}$ and $\beta \in \mathbb{R}^p$ such that $\nabla l(\beta_0, \beta) = 0$:

$$(\beta_0, \beta) \leftarrow (\beta_0, \beta) - \{\nabla^2 l(\beta_0, \beta)\}^{-1} \nabla l(\beta_0, \beta) ,$$

where $\nabla f(v) \in \mathbb{R}^{p+1}$ is a vector such that the i-th element is $\dfrac{\partial f}{\partial v_i}$, and $\nabla^2 f(v) \in \mathbb{R}^{(p+1) \times (p+1)}$ is a square matrix such that the (i, j)-th element is $\dfrac{\partial^2 f}{\partial v_i \partial v_j}$. In the following, for ease of notation, we write $(\beta_0, \beta) \in \mathbb{R} \times \mathbb{R}^p$ as $\beta \in \mathbb{R}^{p+1}$.

If we differentiate the negative log-likelihood $l(\beta_0, \beta)$ and if we let $v_i = e^{-y_i(\beta_0 + x_i \beta)}$, $i = 1, \ldots, N$, the vector $\nabla l(\beta_0, \beta) \in \mathbb{R}^{p+1}$ such that the j-th element is $\dfrac{\partial l(\beta_0, \beta)}{\partial \beta_j}$, $j = 0, 1, \ldots, p$, can be expressed by $\nabla l(\beta_0, \beta) = -X^T u$ with

$$u = \begin{bmatrix} \dfrac{y_1 v_1}{1 + v_1} \\ \vdots \\ \dfrac{y_N v_N}{1 + v_N} \end{bmatrix} ,$$

where β_0 is regarded as the 0th element, and the i-th row of X is $[1, x_i] \in \mathbb{R}^{p+1}$. If we note $y_i = \pm 1$, i.e., $y_i^2 = 1$, the matrix $\nabla^2 l(\beta_0, \beta)$ such that the (j, k)-th element is $\dfrac{\partial^2 l(\beta_0, \beta)}{\partial \beta_j \beta_k}$, $j, k = 0, 1, \ldots, p$, can be expressed by $\nabla^2 l(\beta_0, \beta) = X^T W X$ with

$$W = \begin{bmatrix} \dfrac{v_1}{(1 + v_1)^2} & \cdots & 0 \\ \vdots & \ddots & \vdots \\ 0 & \cdots & \dfrac{v_N}{(1 + v_N)^2} \end{bmatrix} .$$

Using such W and u, the update rule can be written as

$$\beta \leftarrow \beta + (X^T W X)^{-1} X^T u .$$

In addition, if we introduce the variable $z := X\beta + W^{-1}u \in \mathbb{R}^N$, the formula becomes simpler:

$$\beta \leftarrow (X^T W X)^{-1} X^T W z .$$

Example 33 We wrote a Python program that solves $\nabla l(\beta_0, \beta) = 0$ and executed it for the following data.

```python
N=1000; p=2
X=randn(N,p)
X=np.insert(X,0,1,axis=1)
beta=randn(p+1)
y=[]
prob=1/(1+np.exp(X@beta))
for i in range(N):
    if (np.random.rand(1)>prob[i]):
        y.append(1)
    else :
        y.append(-1)
# Data generation ends here
beta # check
```

```
array([ 0.79985659, -1.31770628, -0.23553563])
```

```python
# likelihood estimation
beta=np.inf
gamma=randn(p+1) # initial value of beta
print (gamma)
while (np.sum((beta-gamma)**2)>0.001):
    beta=gamma
    s=X@beta
    v=np.exp(-s*y)
    u=(y*v)/(1+v)
    w=v/((1+v)**2)
    W=np.diag(w)
    z=s+u/w
    gamma=np.linalg.inv(X.T@W@X)@X.T@W@z
    print (gamma)
```

```
[-1.00560507  0.44039528 -0.89669456]
[ 1.73215544 -1.89462271  1.11707796]
[-0.25983643 -0.38933759 -1.10645012]
[ 0.81463839 -1.04443553  0.39176123]
[ 0.7458049  -1.3256336  -0.08413818]
[ 0.79163801 -1.41592785 -0.09332545]
[ 0.7937899  -1.4203184  -0.09373029]
```

We found that the results were almost correct.

For some cases, the maximum likelihood solution cannot be obtained even if we apply the Newton–Raphson method. For example, if the observations satisfy

$y_i(\beta_0 + x_i\beta) \geq 0$, $(x_i, y_i) \in \mathbb{R}^p \times \mathbb{R}$, $i = 1, \ldots, N$, then the maximum likelihood estimate of logistic regression cannot be obtained. In fact, the terms in the exponent part of

$$\prod_{i=1}^{N} \frac{1}{1 + \exp\{-y_i(\beta_0 + x_i\beta)\}}$$

can be all negative, which means that the exponent can diverge to $-\infty$ if we multiply β_0 and β by 2. Thus, the likelihood can approach one by choosing some β_0 and β. Even if we do not meet such conditions, if p is large compared to N, the possibility of the parameter being infinitely large increases.

Example 34 For $p = 1$, we estimated the coefficients $\hat{\beta}_0$, $\hat{\beta}_1$ of logistic regression using the training data with $N/2$ samples and predicted the response of the covariate values in the $N/2$ test data.

```
# data genetration
n=100
x=np.concatenate([randn(n)+1,randn(n)-1],0)
y=np.concatenate([np.ones(n),-np.ones(n)],0)
train=np.random.choice(2*n,int(n),replace=False) #  indices for training
    data
test=list(set(range(2*n))-set(train))      # indices for test data
X=np.insert(x[train].reshape(-1,1),0,1,axis=1)
Y=y[train]
# All 1 columns are added  to the left of x
```

```
# The value may not converge with some  initial value of gamma , so we may
    perform several times.
p=1
beta=[0,0]; gamma=randn(p+1)
print (gamma)
while (np.sum((beta-gamma)**2)>0.001):
    beta=gamma
    s=X@beta
    v=np.exp(-s*Y)
    u=(Y*v)/(1+v)
    w=v/((1+v)**2)
    W=np.diag(w)
    z=s+u/w
    gamma=np.linalg.inv(X.T@W@X)@X.T@W@z
    print (gamma)
```

```
[0.20382031 0.19804102]
[0.17521272 1.13479347]
[0.29020473 1.72206578]
[0.38156063 2.04529677]
[0.40773631 2.1233337 ]
[0.40906736 2.12699164]
```

```
def table_count(m,u,v):
    n=u.shape[0]
    count=np.zeros([m,m])
    for i in range(n):
```

```
        count[int(u[i]),int(v[i])]+=1
    return (count)
```

```
ans=y[test]  # answer
pred=np.sign(gamma[0]+x[test]*gamma[1])   # predicted value
ans=(ans+1)/2      # Change from -1,1 to 0,1.
pred=(pred+1)/2    # Change from -1,1 to 0,1.
table_count(3,ans, pred)
```

```
array([[41.,  9.],
       [ 5., 45.]])
```

We set up a data frame with the pairs of covariate and response values and divided the $N = 2n$ data into training and test sets of size n. The finally obtained values of y are the correct values, and we predicted each of the y values based on the estimates of β_0 and β_1 and whether each of the zs is positive or negative. The table expresses the numbers of correct and incorrect answers, and the correct rate in this experiment was $(41 + 45)/100 = 0.86$.

3.3 Linear and Quadratic Discrimination

As before, we find the map $x \in \mathbb{R}^p \mapsto y \in \{-1, 1\}$ to minimize the error probability, given the observations $x_1, \ldots, x_N \in \mathbb{R}^p$, $y_1, \ldots, y_N \in \{-1, 1\}$. In this section, we assume that the distributions of $x \in \mathbb{R}^p$ given $y = \pm 1$ are $N(\mu_{\pm 1}, \Sigma_{\pm 1})$ and write the probability density functions by

$$f_{\pm 1}(x) = \frac{1}{\sqrt{(2\pi)^p \det \Sigma}} \exp\left\{-\frac{1}{2}(x - \mu_{\pm 1})^T \Sigma_{\pm 1}^{-1}(x - \mu_{\pm 1})\right\}. \tag{3.3}$$

In addition, we introduce the notion of prior probabilities of events: we assume that the probabilities of responses $y = \pm 1$ are known before seeing the covariates x, which we term the prior probability. For example, we may estimate the probability of the response being $\pi_{\pm 1}$ from the ratio of the two from $y_1, \ldots, y_N$ in the training data. On the other hand, we refer to

$$\frac{\pi_{\pm 1} f_{\pm 1}(x)}{\pi_1 f_1(x) + \pi_{-1} f_{-1}(x)}$$

as the posterior probability of $y = \pm 1$ given x. We can minimize the error probability by estimating $y = 1$ if

$$\frac{\pi_1 f_1(x)}{\pi_1 f_1(x) + \pi_{-1} f_{-1}(x)} \geq \frac{\pi_{-1} f_{-1}(x)}{\pi_1 f_1(x) + \pi_{-1} f_{-1}(x)},$$

which is equivalent to

$$\pi_1 f_1(x) \geq \pi_{-1} f_{-1}(x) , \tag{3.4}$$

and $y = -1$ otherwise. The procedure assumes that $f_{\pm 1}$ follows a Gaussian distribution and that the expectation $\mu_{\pm 1}$ and covariance matrix $\Sigma_{\pm 1}$ are known, and that $\pi_{\pm 1}$ is known. For actual situations, we need to estimate these entities from the training data.

The principle of maximizing the posterior probability is applied not only to the binary case ($K = 2$) but also to the general case $K \geq 2$, where K is the number of values that the response takes. The probability that response $y = k$ given covariates x is $P(y = k|x)$ for $k = 1, \ldots, K$. If we estimate $y = \hat{k}$, then the probability of the estimate being correct is $1 - \sum_{k \neq \hat{k}} P(y = k|x) = P(y = \hat{k}|x)$. Thus, choosing a k that maximizes the posterior probability $P(y = \hat{k}|x)$ as $\hat{k}$ minimizes the average error probability when the prior probability is known.

In the following, assuming $K = 2$ for simplicity, we see the properties at the border between $y = \pm 1$ when we maximize the posterior probability:

$$-(x - \mu_1)^T \Sigma_1^{-1}(x - \mu_1) + (x - \mu_{-1})^T \Sigma_{-1}^{-1}(x - \mu_{-1}) = \log \frac{\det \Sigma_1}{\det \Sigma_{-1}} - 2 \log \frac{\pi_1}{\pi_{-1}} ,$$

where the equation is obtained from (3.3) and (3.4). In general, the border is a function of the quadratic forms $x^T \Sigma_1^{-1} x$ and $x^T \Sigma_{-1}^{-1} x$ of x (quadratic discrimination).

In particular, when $\Sigma_1 = \Sigma_{-1}$, if we write them as Σ, the border becomes a surface (a line when $p = 2$), which we call linear discrimination. In fact, the terms $x^T \Sigma_1^{-1} x = x^T \Sigma_{-1}^{-1} x$ are canceled out, and the border becomes

$$2(\mu_1 - \mu_{-1})^T \Sigma^{-1} x - (\mu_1^T \Sigma^{-1} \mu_1 - \mu_{-1}^T \Sigma^{-1} \mu_{-1}) = -2 \log \frac{\pi_1}{\pi_{-1}} ,$$

or more simply,

$$(\mu_1 - \mu_{-1})^T \Sigma^{-1}(x - \frac{\mu_1 + \mu_{-1}}{2}) = -\log \frac{\pi_1}{\pi_{-1}} .$$

Thus, if $\pi_1 = \pi_{-1}$, then the border is $x = \dfrac{\mu_1 + \mu_{-1}}{2}$.

If $\pi_{\pm 1}$ and $f_{\pm 1}$ are unknown, we need to estimate them from the training data.

Example 35 For artificially generated data, we estimated the averages and covariances of covariates x for a response $y = \pm 1$, and drew the border.

```
# True parameters
mu_1=np.array([2,2]); sigma_1=2; sigma_2=2; rho_1=0
mu_2=np.array([-3,-3]); sigma_3=1; sigma_4=1; rho_2=-0.8
```

```
# generate data based on true parameters
n=100
u=randn(n); v=randn(n)
x_1=sigma_1*u+mu_1[0]; y_1=(rho_1*u+np.sqrt(1-rho_1**2)*v)*sigma_2+mu_1[1]
u=randn(n); v=randn(n)
x_2=sigma_3*u+mu_2[0]; y_2=(rho_2*u+np.sqrt(1-rho_2**2)*v)*sigma_4+mu_2[1]

# estimate the parameters from the data
mu_1=np.average((x_1,y_1),1); mu_2=np.average((x_2,y_2),1)
df=np.array([x_1,y_1]); mat=np.cov(df,rowvar=1); inv_1=np.linalg.inv(mat);
   de_1=np.linalg.det(mat)   #
df=np.array([x_2,y_2]); mat=np.cov(df,rowvar=1); inv_2=np.linalg.inv(mat);
   de_2=np.linalg.det(mat)   #
```

```
# substitute the parameters into the distribution formula
def f(x,mu,inv,de):
    return(-0.5*(x-mu).T@inv@(x-mu)-0.5*np.log(de))
def f_1(u,v):
    return f(np.array([u,v]),mu_1,inv_1,de_1)
def f_2(u,v):
    return f(np.array([u,v]),mu_2,inv_2,de_2)
```

```
#  generate contour data
# draw a boundary line where this value is 0
pi_1=0.5; pi_2=0.5
u=v=np.linspace(-6,6,50)
m=len(u)
w=np.zeros([m,m])
for i in range(m):
    for j in range(m):
        w[i,j]=np.log(pi_1)+f_1(u[i],v[j])-np.log(pi_2)-f_2(u[i],v[j])

# plotting Boundaries and Data
plt.contour(u,v,w,levels=0,colors=['black'])
plt.scatter(x_1,y_1,c="red")
plt.scatter(x_2,y_2,c="blue")
```

We show the covariates for each response and the generated border in Fig. 3.3 (Right). If the covariance matrices are equal, we change the lines marked with "#" as follows:

```
# Linear Discrimination (Figure 2 . 3 left) (if we  assume the variance is
   equal)
# modify the lines marked with # as follows
xx=np.concatenate((x_1-mu_1[0],x_2-mu_2[0]),0).reshape(-1,1)
yy=np.concatenate((y_1-mu_1[1],y_2-mu_2[1]),0).reshape(-1,1)
df=np.concatenate((xx,yy),1) #  data was merged vertically.
mat=np.cov(df,rowvar=0)          #  rowvar=0 because of the vertical direction
inv_1=np.linalg.inv(mat)
de_1=np.linalg.det(mat)
inv_2=inv_1; de_2=de_1
w=np.zeros([m,m])
for i in range(m):
    for j in range(m):
        w[i,j]=np.log(pi_1)+f_1(u[i],v[j])-np.log(pi_2)-f_2(u[i],v[j])
plt.contour(u,v,w,levels=0,colors=['black'])
plt.scatter(x_1,y_1,c="red")
plt.scatter(x_2,y_2,c="blue")
```

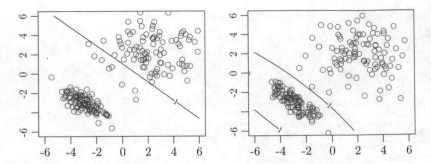

Fig. 3.3 Linear Discrimination (Left) and Quadratic Discrimination (Right): The border is a line if the covariance matrices are equal; otherwise, it is a quadratic (elliptic) curve. In the former case, if the prior probabilities and the covariance matrices are equal, then the border is the vertical bisector of the line connecting the centers

We show the output in Fig. 3.3 (Left).

Example 36 (Fisher's Iris Dataset)   Even when the response takes more than two values, we can choose the response with the maximum posterior probability. Fisher's Iris dataset contains four covariates (the petal length, petal width, sepal length, and sepal width), and the response variable can be three species of irises (*Iris setosa*, *Iris virginica*, and *Iris versicolor*). Each of the three species contains 50 samples ($N = 150$, $p = 4$). We construct the classifier via quadratic discrimination and evaluate it using the test dataset that are different from the training data.

```
from sklearn.datasets import load_iris
```

```
iris=load_iris()
iris.target_names
x=iris.data
y=iris.target
n=len(x)
train=np.random.choice(n,int(n/2),replace=False)
test=list(set(range(n))-set(train))
# estimate parameter
X=x[train,:]
Y=y[train]
mu=[]
covv=[]
for j in range(3):
    xx=X[Y==j,:]
    mu.append(np.mean(xx,0))
    covv.append(np.cov(xx,rowvar=0))
```

```
# Definitions of distributions which we substitute the estimated parameters
def f(w,mu,inv,de):
    return -0.5*(w-mu).T@inv@(w-mu)-0.5*np.log(de)
def g(v,j):
    return f(v,mu[j],np.linalg.inv(covv[j]),np.linalg.det(covv[j]))
```

```
z=[]
for i in test:
    z.append(np.argsort([-g(x[i,],0),-g(x[i,],1),-g(x[i,],2)])[0])
table_count(3,y[test],z)
```

```
array([[27.,  0.,  0.],
       [ 0., 20.,  4.],
       [ 0.,  0., 24.]])
```

If the prior probabilities of the three species are not equal, for example, if those of *Iris setosa*, *Iris virginica*, and *Iris versicolor* are 0.5, 0.25, and 0.25, respectively, then, we add the logarithm of the prior probabilities to the variables a, b, c in the program.

3.4 *k*-Nearest Neighbor Method

The *k*-nearest neighbor method does not require constructing a specific rule from the training data $(x_1, y_1), \ldots, (x_N, y_N) \in \mathbb{R}^p \times$(finite set). Suppose that given new data $x_* \in \mathbb{R}^p$, $x_i, i \in S$, are the k training data such that the distances between x_i and x_* are the smallest, where S is a subset of $\{1, \cdots, n\}$ of size k. The *k*-nearest neighbor method predicts the response y_* of x_* by the majority of $y_i, i \in S$.).

For example, suppose that $N = 5$, $p = 1$ and that the data are given as below. If $k = 3$ and $x_* = 1.6$, then $S = \{3, 4, 5\}$ and the majority class is $y_* = 0$. If $k = 2$ and $x_* = -2.2$, then $S = \{1, 2\}$. However, in that case, the majority is not unique. Then, we remove one element from S. Because $x_1 = -2.1$ is close to $x_* = -2.2$, we set $S = \{2\}$ and $y_* = -1$.

x_i	−2.1	−3.7	1.3	0.4	1.5
y_i	−1	1	0	0	1

For example, we may construct the following procedure for the *k*-nearest neighbor method that uses a tie-breaking rule, the $m - 1$ responses among the closest $m - 1$ responses are compared when the majority among the closest m responses are not unique:

```
def knn_1(x,y,z,k):
    x=np.array(x); y=np.array(y)
    dis=[]
    for i in range(x.shape[0]):
        dis.append(np.linalg.norm(z-x[i,]))
    S=np.argsort(dis)[0:k]    # k indices which The distance is close
    u=np.bincount(y[S])        # count the number
    m=[i for i, x in enumerate(u) if x==max(u)] # index of high frequent
    # Processing of the brakings (if the frequency is more than 2)
    while (len(m)>1):
        k=k-1
        S=S[0:k]
        u=np.bincount(y[S])
        m=[i for i, x in enumerate(u) if x==max(u)] # index of high frequent
    return m[0]
```

If there is more than one majority class, we remove the $i \in S$ such that the distance between x_j and x_* is the largest among x_j, $j \in S_j$ and continue to find the majority. If S contains exactly one element, eventually, we identify the majority class.

For multiple x_*s, we may extend the above procedure to the following:

```
# generalize
def knn(x,y,z,k):
    w=[]
    for i in range(z.shape[0]):
        w.append(knn_1(x,y,z[i,],k))
    return w
```

We find that the smaller k, the more sensitive the border is to the training data.

Example 37 (Fisher's Iris Dataset)

```
from sklearn.datasets import load_iris
```

```
iris=load_iris()
iris.target_names
x=iris.data
y=iris.target
n=x.shape[0]
train=np.random.choice(n,int(n/2),replace=False)
test=list(set(range(n))-set(train))
w=knn(x[train,],y[train],x[test,],k=3)
table_count(3,y[test],w)
```

```
array([[25.,  0.,  0.],
       [ 0., 26.,  4.],
       [ 0.,  1., 19.]])
```

3.5 ROC Curves

Although maximizing the posterior probability is valid in many cases in the sense of minimizing the error probability, however, we may want to improve an alternative performance even if we lose the merit of minimizing the error probability.

For example, during credit card screening, less than 3% of applicants have problems. In this case, if all the applications are approved, an error rate of 3% is attained. However, in that case, the card company claims that there are risks and rejects at least 10% of the applications.

In cancer screening, although only 3% of people have cancer, more than 20% of people who have been screened are diagnosed with cancer. Considering a sick person as healthy is riskier than treating a healthy person as unhealthy. If a doctor does not want to take responsibility, he may judge more people as having cancer.

In other words, depending on the balance between the risk of mistakenly considering a healthy person as sick (type I error) and the risk of assuming a sick

Table 3.1 Examples of types I and II errors

	Type I Error	Type II Error
Quality control	Identify good products as defective	Identify defective products as good
Medical diagnosis	Identify healthy people as sick	Identify sick people as healthy
Criminal investigation	Identify the criminal as not a criminal	Treat noncriminals as criminals
Entrance exams	Reject excellent students	Allow inferior students to enter

person as healthy (type II error) (Table 3.1), the criterion for judgment differs. In other words, it is necessary to consider the optimality of each of the ways of balancing the risks. We use terms such as true positives, false positives, false negatives, and true negatives as defined below.

	Sick	Healthy
Treating as sick	True Positive	False Positive
Treating as healthy	False Negative	True Negative

The rates of the type I and type II errors are α and β, respectively, the power and false positive rate are defined as follows:

$$\text{Power} = \frac{TP}{TP + FN} = 1 - \beta$$

$$\textit{False Positive Rate} = \frac{FP}{FP + TN} = \alpha.$$

For each false positive rate (α), consider maximizing the power $1 - \beta$ (the Neyman–Pearson criterion). In that case, there are countless ways of testing depending on how to balance the two values. The curve with the false positive rate on the horizontal axis and the power on the vertical axis is called the receiver operating characteristic (ROC) curve. The higher the curve goes to the upper-left corner of the plot, that is, the larger the area under the ROC curve (AUC, maximum of 1), the better the test performs.

Example 38 Let $f_1(x)$ and $f_0(x)$ be the distributions for a measurement x of people with a disease and healthy people, respectively. For each positive θ, the decision was made to determine whether the person had the symptom if

$$\frac{f_1(x)}{f_0(x)} \geq \theta .$$

Fig. 3.4 The ROC curve shows all the performances of the test for acceptable false positives

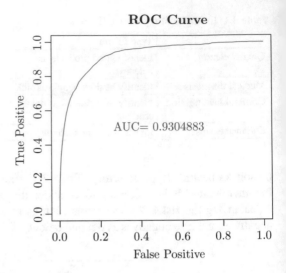

In the following, the distributions of sick and healthy people are $N(1, 1)$ and $N(-1, 1)$, respectively, and the ROC curve is shown in Fig. 3.4:

```
N_0=10000; N_1=1000
mu_1=1; mu_0=-1  # Sick  :1 , Normal :0
var_1=1; var_0=1
x=np.random.normal(mu_0,var_0,N_0)
y=np.random.normal(mu_1,var_1,N_1)
theta_seq=np.exp(np.arange(-10,100,0.1))
U=[]; V=[]
for i in range(len(theta_seq)):
    u=np.sum((stats.norm.pdf(x,mu_1,var_1)/stats.norm.pdf(x,mu_0,var_0))>
        theta_seq[i])/N_0
    # Treat a person who are not sick as sick
    v=np.sum((stats.norm.pdf(y,mu_1,var_1)/stats.norm.pdf(y,mu_0,var_0))>
        theta_seq[i])/N_1
    # Treating a sick person as sick
    U.append(u); V.append(v)
```

```
AUC=0 # estimate the  area
for i in range(len(theta_seq)-1):
    AUC=AUC+np.abs(U[i+1]-U[i])*V[i]
```

```
plt.plot(U,V)
plt.xlabel("False_Positive")
plt.ylabel("True_Positive")
plt.title("ROC_curve")
plt.text(0.3,0.5,'AUC={}'.format(AUC),fontsize=15)
```

```
Text(0.3, 0.5, 'AUC=0.9301908000000001')
```

Exercises 19–31

19. We assume that there exist $\beta_0 \in \mathbb{R}$ and $\beta \in \mathbb{R}^p$ such that for $x \in \mathbb{R}^p$, the probabilities of $Y = 1$ and $Y = -1$ are $\dfrac{e^{\beta_0 + x\beta}}{1 + e^{\beta_0 + x\beta}}$ and $\dfrac{1}{1 + e^{\beta_0 + x\beta}}$, respectively. Show that the probability of $Y = y \in \{-1, 1\}$ can be written as $\dfrac{1}{1 + e^{-y(\beta_0 + x\beta)}}$.

20. For $p = 1$ and $\beta > 0$, show that the function $f(x) = \dfrac{1}{1 + e^{-(\beta_0 + x\beta)}}$ is monotonically increasing for $x \in \mathbb{R}$ and convex and concave in $x < -\beta_0/\beta$ and $x > -\beta_0/\beta$, respectively. How does the function change as β increases? Execute the following to answer this question:

```
def f(x):
    return np.exp(beta_0+beta*x)/(1+np.exp(beta_0+beta*x))
```

```
beta_0=0
beta_seq=np.array([0,0.2,0.5,1,2,10])
x_seq=np.arange(-10,10,0.1)
plt.xlabel("x")
plt.ylabel("P(Y=1|x)")
plt.title("logistic_curve")
for i in range(beta_seq.shape[0]):
    beta=beta_seq[i]
    p=f(x_seq)
    plt.plot(x_seq,p,label='{}'.format(beta))
plt.legend(loc='upper_left')
```

21. We wish to obtain the estimates of $\beta_0 \in \mathbb{R}$ and $\beta \in \mathbb{R}^p$ by maximizing the likelihood $\displaystyle\prod_{i=1}^{N} \dfrac{1}{1 + e^{-y_i(\beta_0 + x_i\beta)}}$, or equivalently, by minimizing the negated logarithm

$$l(\beta_0, \beta) = \sum_{i=1}^{N} \log(1 + v_i), \quad v_i = e^{-y_i(\beta_0 + x_i\beta)}$$

from observations $(x_1, y_1), \ldots, (x_N, y_N) \in \mathbb{R}^p \times \{-1, 1\}$ (maximum likelihood). Show that $l(\beta_0, \beta)$ is convex by obtaining the derivative $\nabla l(\beta_0, \beta)$ and the second derivative $\nabla^2 l(\beta_0, \beta)$. Hint: Let $\nabla l(\beta_0, \beta)$ and $\nabla^2 l(\beta_0, \beta)$ be the column vector of size $p + 1$ such that the j-th element is $\dfrac{\partial l}{\partial \beta_j}$ and the matrix of size $(p + 1) \times (p + 1)$ such that the (j, k)-th element is $\dfrac{\partial^2 l}{\partial \beta_j \partial \beta_k}$, respectively. Simply show that the matrix is nonnegative definite. To this end, show that $\nabla^2 l(\beta_0, \beta) = X^T W X$. If W is diagonal, then it can be written as $W = U^T U$, where the diagonal elements of U are the square roots of W, which means $\nabla^2 l(\beta_0, \beta) = (UX)^T UX$.

22. Solve the following equations via the Newton–Raphson method by constructing a Python program:

(a) For $f(x) = x^2 - 1$, set $x = 2$ and repeat the recursion $x \leftarrow x - f(x)/f'(x)$ 100 times.

(b) For $f(x, y) = x^2 + y^2 - 1$, $g(x, y) = x + y$, set $(x, y) = (1, 2)$ and repeat the recursion 100 times.

$$
\begin{bmatrix} x \\ y \end{bmatrix} \leftarrow \begin{bmatrix} x \\ y \end{bmatrix} - \begin{bmatrix} \dfrac{\partial f(x, y)}{\partial x} & \dfrac{\partial f(x, y)}{\partial y} \\ \dfrac{\partial g(x, y)}{\partial x} & \dfrac{\partial g(x, y)}{\partial y} \end{bmatrix}^{-1} \begin{bmatrix} f(x, y) \\ g(x, y) \end{bmatrix}
$$

Hint: Define the procedure and repeat it one hundred times.

```
def f(z):
    return z[0]**2+z[1]**2-1
def dfx(z):
    return 2*z[0]
def dfy(z):
    return 2*z[1]
def g(z):
    return z[0]+z[1]
def dgx(z):
    return 1
def dgy(z):
    return 1
z=np.array([1,2]) # initial value
```

23. We wish to solve $\nabla l(\beta_0, \beta) = 0$, $(\beta_0, \beta) \in \mathbb{R} \times \mathbb{R}^p$ in Problem 21 via the Newton–Raphson method using the recursion

$$
(\beta_0, \beta) \leftarrow (\beta_0, \beta) - \{\nabla^2 l(\beta_0, \beta)\}^{-1} \nabla l(\beta_0, \beta) ,
$$

where $\nabla f(v) \in \mathbb{R}^{p+1}$ and $\nabla^2 f(v) \in \mathbb{R}^{(p+1)\times(p+1)}$ are the vector such that the i-th element is $\dfrac{\partial f}{\partial v_i}$ and the square matrix such that the (i, j)-th element is $\dfrac{\partial^2 f}{\partial v_i \partial v_j}$, respectively. In the following, for ease of notation, we write $(\beta_0, \beta) \in \mathbb{R} \times \mathbb{R}^p$ by $\beta \in \mathbb{R}^{p+1}$. Show that the update rule can be written as

$$
\beta_{new} \leftarrow (X^T W X)^{-1} X^T W z , \tag{3.5}
$$

where $u \in \mathbb{R}^{p+1}$ such that $\nabla l(\beta_{old}) = -X^T u$ and $W \in \mathbb{R}^{(p+1)\times(p+1)}$ such that $\nabla^2 l(\beta_{old}) = X^T W X$, $z \in \mathbb{R}$ is defined by $z := X\beta_{old} + W^{-1}u$, and $X^T W X$ is assumed to be nonsingular. Hint: The update rule can be written as $\beta_{new} \leftarrow \beta_{old} + (X^T W X)^{-1} X^T u$.

24. We construct a procedure to solve Problem 23. Fill in blanks (1)(2)(3), and
    examine that the procedure works.

```
N=1000; p=2
X=randn(N,p)
X=np.insert(X,0,1,axis=1)
beta=randn(p+1)
y=[]
prob=1/(1+np.exp(X@beta))
for i in range(N):
    if (np.random.rand(1)>prob[i]):
        y.append(1)
    else :
        y.append(-1)
# # Data generation ends here
beta # check
```

```
array([ 0.79985659, -1.31770628, -0.23553563])
```

```
# # likelihood estimation
beta=np.inf
gamma=randn(p+1) #
print (gamma)
while (np.sum((beta-gamma)**2)>0.001):
    beta=gamma
    s=X@beta
    v=np.exp(-s*y)
    u=# blank(1) #
    w=# blank(2) #
    W=np.diag(w)
    z=# blank(3) #
    gamma=np.linalg.inv(X.T@W@X)@X.T@W@z
    print (gamma)
```

25. If the condition $y_i(\beta_0 + x_i\beta) \geq 0$, $(x_i, y_i) \in \mathbb{R}^P \times \mathbb{R}$, $i = 1, \ldots, N$ is met,
    we cannot obtain the parameters of logistic regression via maximum likelihood.
    Why?

26. For $p = 1$, we wish to estimate the parameters of logistic regression from $N/2$
    training data and to predict the responses of the $N/2$ test data that are not used
    as the training data. Fill in the blanks and execute the program.

```
# data genetration
n=100
x=np.concatenate([randn(n)+1,randn(n)-1],0)
y=np.concatenate([np.ones(n),-np.ones(n)],0)
train=np.random.choice(2*n,int(n),replace=False) # indices for training
    data
test=list(set(range(2*n))-set(train))       # indices for test data
X=np.insert(x[train].reshape(-1,1), 0, 1, axis=1)
Y=y[train]
# All 1 columns are added to the left of x
```

```
# The value may not converge with some initial value of gamma, so we may
    perform severaltimes.
p=1
beta=[0,0]; gamma=randn(p+1)
```

```
print (gamma)
while (np.sum((beta-gamma)**2)>0.001):
    beta=gamma
    s=X@beta
    v=np.exp(-s*Y)
    u=(Y*v)/(1+v)
    w=v/((1+v)**2)
    W=np.diag(w)
    z=s+u/w
    gamma=np.linalg.inv(X.T@W@X)@X.T@W@z
    print (gamma)
```

```
def table_count(m,u,v):
    n=u.shape[0]
    count=np.zeros([m,m])
    for i in range(n):
        # blank(1) #+=1
    return (count)
```

```
ans=y[test]  # answer
pred=# blank(2) #
ans=(ans+1)/2      # change from -1,1, to 0,1.
pred=(pred+1)/2    # change from -1,1, to 0,1.
table_count(3,ans,pred)
```

Hint: For prediction, see whether $\beta_0 + x\beta_1$ is positive or negative.

27. In linear discrimination, let π_k be the prior probability of $Y = k$ for $k = 1, \ldots, m$ ($m \geq 2$), and let $f_k(x)$ be the probability density function of the p covariates $x \in \mathbb{R}^p$ given response $Y = k$ with mean $\mu_k \in \mathbb{R}^p$ and covariance matrix $\Sigma_k \in \mathbb{R}^{p \times p}$. We consider the set $S_{k,l}$ of $x \in \mathbb{R}^p$ such that

$$\frac{\pi_k f_k(x)}{\displaystyle\sum_{j=1}^{K} \pi_j f_j(x)} = \frac{\pi_l f_l(x)}{\displaystyle\sum_{j=1}^{K} \pi_j f_j(x)}$$

for $k, l = 1, \ldots, m, k \neq l$.

(a) Show that when $\pi_k = \pi_l$, $S_{k,l}$ is the set of $x \in \mathbb{R}^p$ on the quadratic surface

$$-(x - \mu_k)^T \Sigma_k^{-1}(x - \mu_k) + (x - \mu_l)^T \Sigma_l^{-1}(x - \mu_l) = \log \frac{\det \Sigma_k}{\det \Sigma_l} .$$

(b) Show that when $\Sigma_k = \Sigma_l (= \Sigma)$, $S_{k,l}$ is the set of $x \in \mathbb{R}^p$ on the surface $a^T x + b = 0$ with $a \in \mathbb{R}^p$ and $b \in \mathbb{R}$ and express a, b using $\mu_k, \mu_l, \Sigma, \pi_k, \pi_l$.

(c) When $\pi_k = \pi_l$ and $\Sigma_k = \Sigma_l$, show that the surface of (b) is $x = (\mu_k + \mu_l)/2$.

28. In the following, we wish to estimate distributions from two classes and draw a boundary line that determines the maximum posterior probability. If the covariance matrices are assumed to be equal, how do the boundaries change? Modify the program.

```
# True parameters
mu_1=np.array([2,2]); sigma_1=2; sigma_2=2; rho_1=0
mu_2=np.array([-3,-3]); sigma_3=1; sigma_4=1; rho_2=-0.8

# generate data based on true parameter
n=100
u=randn(n); v=randn(n)
x_1=sigma_1*u+mu_1[0]; y_1=(rho_1*u+np.sqrt(1-rho_1**2)*v)*sigma_2+mu_1[1]
u=randn(n); v=randn(n)
x_2=sigma_3*u+mu_2[0]; y_2=(rho_2*u+np.sqrt(1-rho_2**2)*v)*sigma_4+mu_2[1]

# estimate the parameters from the data
mu_1=np.average((x_1,y_1),1); mu_2=np.average((x_2,y_2),1)
df=np.array([x_1,y_1]); mat=np.cov(df,rowvar=1); inv_1=np.linalg.inv(mat);
    de_1=np.linalg.det(mat)  #
df=np.array([x_2,y_2]); mat=np.cov(df,rowvar=1); inv_2=np.linalg.inv(mat);
    de_2=np.linalg.det(mat)  #
```

```
# substitute the parameters into the distribution formula
def f(x,mu,inv,de):
    return(-0.5*(x-mu).T@inv@(x-mu)-0.5*np.log(de))
def f_1(u,v):
    return f(np.array([u,v]),mu_1,inv_1,de_1)
def f_2(u,v):
    return f(np.array([u,v]),mu_2,inv_2,de_2)
```

```
# generate contour data
# draw a boundary line where this value is 0
pi_1=0.5; pi_2=0.5
u=v=np.linspace(-6,6,50)
m=len(u)
w=np.zeros([m,m])
for i in range(m):
    for j in range(m):
        w[i,j]=np.log(pi_1)+f_1(u[i],v[j])-np.log(pi_2)-f_2(u[i],v[j])
# plotting Boundaries and Data
plt.contour(u,v,w,levels=1,colors=['black'])
plt.scatter(x_1,y_1,c="red")
plt.scatter(x_2,y_2,c="blue")
```

Hint: Modify the lines marked with #.

29. Even in the case of three or more values, we can select the class that maximizes the posterior probability. From four covariates (length of sepals, width of sepals, length of petals, and width of petals) of Fisher's iris data, we wish to identify the three types of irises (Setosa, Versicolor, and Virginica) via quadratic discrimination. Specifically, we learn rules from training data and evaluate them with test data. Assuming $N = 150$ and $p = 4$, each of the three irises contains 50 samples, and the prior probability is expected to be equal to $1/3$. If we find that the prior probabilities of Setosa, Versicolor, and Virginica irises are 0.5, 0.25, 0.25, how should the program be changed to determine the maximum posterior probability?

```
from sklearn.datasets import load_iris
```

```
iris=load_iris()
iris.target_names
x=iris.data
y=iris.target
n=len(x)
train=np.random.choice(n,int(n/2),replace=False)
test=list(set(range(n))-set(train))
# estimate parameter
X=x[train,:]
Y=y[train]
mu=[]
covv=[]
for j in range(3):
    xx=X[Y==j,:]
    mu.append(np.mean(xx,0))
    covv.append(np.cov(xx,rowvar=0))
```

```
# Definitions of distributions which we substitute the estimated parameters
def f(w,mu,inv,de):
    return -0.5*(w-mu).T@inv@(w-mu)-0.5*np.log(de)
def g(v,j):
    return f(v,mu[j],np.linalg.inv(covv[j]),np.linalg.det(covv[j]))
```

```
z=[]
for i in test:
    a=g(x[i,],0); b=g(x[i,],1); c=g(x[i,],2)
    if a<b:
        if b<c:
            z.append(2)
        else:
            z.append(1)
    else:
        z.append(0)
u=y[test]
count=np.zeros([3,3])
for i in range(int(n/2)):
    count[u[i],z[i]]+=1
count
```

30. In the k-nearest neighbor method, we do not construct a specific rule from training data $(x_1, y_1), \ldots, (x_N, y_N) \in \mathbb{R}^p \times$(finite set). Suppose that given a new data x_*, x_i, $i \in S$ are the k training data such that the distances between x_i and x_* are the smallest, where S is a subset of $\{1, \ldots, n\}$ of size k. The k-nearest neighbor method predicts the response y_* of x_* by majority voting of y_i, $i \in S$. If there is more than one majority, we remove the $i \in S$ such that the distance between x_j and x_* is the largest among x_j, $j \in S_j$ and continue to find the majority. If S contains exactly one element, we obtain the majority. The following process assumes that there is one test data, but the method can be extended to cases where there is more than one test data. Then, apply the method to the data in Problem 29.

```
def knn_1(x,y,z,k):
    x=np.array(x); y=np.array(y)
    dis=[]
    for i in range(x.shape[0]):
        dis.append(np.linalg.norm(z-x[i,],ord=2))
    S=np.argsort(dis)[0:k]   # k indices which The distance is close
    u=np.bincount(y[S])      # count the number
    m=[i for i, x in enumerate(u) if x==max(u)] # index of high frequent
    # Processing of the brakings (if the frequency is more than 2)
    while (len(m)>1):
        k=k-1
        S=S[0:k]
        u=np.bincount(y[S])
        m=[i for i, x in enumerate(u) if x==max(u)] # index of high
            frequent
    return m[0]
```

31. Let $f_1(x)$ and $f_0(x)$ be the distributions for a measurement x of people with a disease and those without the disease, respectively. For each positive θ, the decision that the person had the symptoms was determined according to whether

$$\frac{f_1(x)}{f_0(x)} \geq \theta .$$

In the following, we suppose that the distributions of sick and healthy people are $N(1, 1)$ and $N(-1, 1)$, respectively. Fill in the blank and draw the ROC curve.

```
N_0=10000; N_1=1000
mu_1=1; mu_0=-1  # Sick :1, Normal :0
var_1=1; var_0=1
x=np.random.normal(mu_0,var_0,N_0)
y=np.random.normal(mu_1,var_1,N_1)
theta_seq=np.exp(np.arange(-10,100,0.1))
U=[]; V=[]
for i in range(len(theta_seq)):
    u=np.sum((stats.norm.pdf(x,mu_1,var_1)/stats.norm.pdf(x,mu_0,var_0))>
        theta_seq[i])/N_0 # Treat a person who are not sick as sick
    v= ## blank ##
    U.append(u); V.append(v)
```

```
AUC=0 # estimate the are
for i in range(len(theta_seq)-1):
    AUC=AUC+np.abs(U[i+1]-U[i])*V[i]
```

```
plt.plot(U,V)
plt.xlabel("False_Positive")
plt.ylabel("True_Positive")
plt.title("ROC_curve")
plt.text(0.3,0.5,'AUC={}'.format(AUC),fontsize=15)
```

Chapter 4
Resampling

Abstract Generally, there is not only one statistical model that explains a phenomenon. In that case, the more complicated the model, the easier it is for the statistical model to fit the data. However, we do not know whether the estimation result shows a satisfactory (prediction) performance for new data different from those used for the estimation. For example, in the forecasting of stock prices, even if the price movements up to yesterday are analyzed so that the error fluctuations are reduced, the analysis is not meaningful if no suggestion about stock price movements for tomorrow is given. In this book, choosing a more complex model than a true statistical model is referred to as overfitting. The term overfitting is commonly used in data science and machine learning. However, the definition may differ depending on the situation, so the author felt that uniformity was necessary. In this chapter, we will first learn about cross-validation, a method of evaluating learning performance without being affected by overfitting. Furthermore, the data used for learning are randomly selected, and even if the data follow the same distribution, the learning result may be significantly different. In some cases, the confidence and the variance of the estimated value can be evaluated, as in the case of linear regression. In this chapter, we will continue to learn how to assess the dispersion of learning results, called bootstrapping.

4.1 Cross-Validation

As we attempted in the previous chapter, it makes sense to remove some of the N tuples for test instead of using them all for estimation. However, in that case, the samples used for estimation are reduced, and a problem occurs, such as the estimation accuracy deteriorating.

Therefore, a method called (k-fold) cross-validation (CV) was devised. Assuming that k is an integer that divides N, $1/k$ of the data are used for the test, and the other $1 - 1/k$ of the data are used to estimate the model; we change the test data and estimate the model k times and evaluate it by the average (Table 4.1). The

© The Author(s), under exclusive license to Springer Nature Singapore Pte Ltd. 2021　　　77
J. Suzuki, *Statistical Learning with Math and Python*,
https://doi.org/10.1007/978-981-15-7877-9_4

Table 4.1 Rotation in the cross-validation approach. Each group consists of N/k samples, which are divided into k groups based on the sample ID: $1 \sim \frac{N}{k}, \frac{N}{k} + 1 \sim \frac{2N}{k}, \dots, (k-2)\frac{N}{k} + 1 \sim (k-1)\frac{N}{k}, (k-1)\frac{N}{k} + 1 \sim N$

	Group 1	Group 2	$\cdots$	Group $k-1$	Group k
First	Test	Estimate	$\cdots$	Estimate	Estimate
Second	Estimate	Test	$\cdots$	Estimate	Estimate
	$\vdots$	$\vdots$	$\ddots$	$\vdots$	$\vdots$
$(k-1)$-th	Estimate	Estimate	$\cdots$	Test	Estimate
k-th	Estimate	Estimate	$\cdots$	Estimate	Test

process of evaluating the prediction error of linear regression is as follows (function `cv_linear`).

```
def cv_linear(X,y,K):
    n=len(y); m=int(n/K)
    S=0
    for j in range(K):
        test=list(range(j*m,(j+1)*m)) # indices for test data
        train=list(set(range(n))-set(test))    # indices for train data
        beta=np.linalg.inv(X[train,].T@X[train,])@X[train,].T@y[train]
        e=y[test]-X[test,]@beta
        S=S+np.linalg.norm(e)**2
    return S/n
```

Example 39 We analyzed the variable selection results of the 10-fold cross-validation approach for linear regression. We assumed that the response depends only on X_3, X_4, X_5 (and the intercept) among the $p = 5$ covariates X_1, X_2, X_3, X_4, X_5.

```
n=100; p=5
X=randn(n,p)
X=np.insert(X,0,1,axis=1)
beta=randn(p+1)
beta[[1,2]]=0
y=X@beta+randn(n)
cv_linear(X[:,[0,3,4,5]],y,10)
```

```
1.1001140673920566
```

```
cv_linear(X,y,10)
```

```
1.156169036077035
```

We evaluated the prediction error via cross-validation for the three cases such that the response depends on $\{X_3, X_4, X_5\}$ and $\{X_1, X_2, X_3, X_4, X_5\}$, respectively. In the first and last cases, it is difficult to see the difference without repeating it. Therefore, the difference was compared 100 times (Fig. 4.1).

Fig. 4.1 The horizontal axis is the prediction error when considering variable selection (some variables are not selected), and this case has a slightly smaller prediction error compared to the case where all the variables are selected (vertical axis)

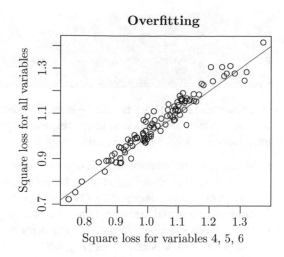

```
n=100; p=5
X=randn(n,p)
X=np.insert(X,0,1,axis=1)
beta=randn(p+1); beta[[1,2]]=0
U=[]; V=[]
for j in range(100):
    y=X@beta+randn(n)
    U.append(cv_linear(X[:,[0,3,4,5]],y,10))
    V.append(cv_linear(X,y,10))
x_seq=np.linspace(0.7,1.5,100)
y=x_seq
plt.plot(x_seq,y,c="red")
plt.scatter(U,V)
plt.xlabel("The_squared_error_in_selecting_variables_4,_5_and_6")
plt.ylabel("The_squared_error_in_selecting_all_variables")
plt.title("Over_fitting_by_selecting_too_many_variables")
```

```
Text(0.5, 1.0, 'Over fitting by selecting too many variables')
```

As seen from Fig. 4.1, overfitting occurs when all the variables are used without performing variable selection.

Example 40 What k of k-fold CV is the best w.r.t. the prediction error depends on the data. Some believe that $k = N$ is optimal. Additionally, $k = 10$ is often used, but there is no theoretical basis for this choice. We generated N datasets ten times and showed how the value changes depending on the value of k (see Fig. 4.2). It seems that the value of k should be at least ten.

```
n=100; p=5
plt.ylim(0.3,1.5)
plt.xlabel("k")
plt.ylabel("the_values_of_CV")
for j in range(2,11,1):
    X=randn(n,p)
    X=np.insert(X,0,1,axis=1)
    beta=randn(p+1)
```

```
y=X@beta+randn(n)
U=[]; V=[]
for k in range(2,n+1,1):
    if n%k==0:
        U.append(k); V.append(cv_linear(X,y,k))
plt.plot(U,V)
```

Cross-validation (CV) is widely used in practical aspects of data science, as well as in variable selection in linear regression problems (e.g., 39). The problems introduced in Chap. 3 are as follows:

- In variable selection for logistic regression, the error rate is compared by CV to find the optimal combination of variables whose coefficients are not zero.
- The error rate is compared by CV between linear discrimination and quadratic discrimination, and the better method is selected.
- In the k-nearest neighbor method, the error rate of each k is compared by CV, and the optimal k is calculated (Fig. 4.3).

Fig. 4.2 How the prediction error of CV changes with k. We simulated artificial data following the same distribution ten times. It can be confirmed that a specific k is not small

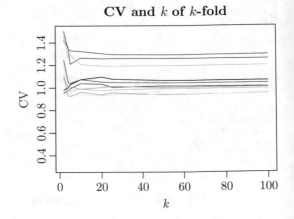

Fig. 4.3 10-fold cross-validation. We show how the error rate of the prediction changes with k in the k-nearest neighbor method

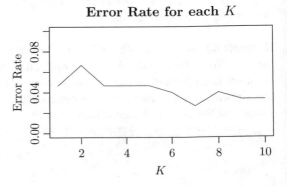

There are numerous ways to prevent overfitting other than CV (described later); however, CV is more advantageous in the sense that they can be applied more generally than CV.

Example 41 With 10-fold cross-validation, we evaluated how the error rate changes for each k in the k-nearest neighbor method for Fisher's Iris dataset. The execution may take more than 10 min (we used the function knn defined in Chap. 3).

```python
from sklearn.datasets import load_iris
```

```python
iris=load_iris()
iris.target_names
x=iris.data
y=iris.target
n=x.shape[0]
index=np.random.choice(n,n,replace=False) # rearrange
x=x[index,]
y=y[index]
```

```python
U=[]
V=[]
top_seq=list(range(0,150,15))
for k in range(1,11,1):
    S=0
    for top in top_seq:
        test=list(range(top,top+15))
        train=list(set(range(150))-set(test))
        knn_ans=knn(x[train,],y[train],x[test,],k=k)
        ans=y[test]
        S=S+np.sum(knn_ans!=ans)
    S=S/n
    U.append(k)
    V.append(S)
plt.plot(U,V)
plt.xlabel("K")
plt.ylabel("error_rate_")
plt.title("_Assessment_of_error_rate_by_CV_")
```

```
Text(0.5, 1.0, ' Assessment of error rate by CV ')
```

4.2 CV Formula for Linear Regression

In the case of k-fold CV, especially when $k = N$ (LOOCV, leave-one-out cross-validation), it takes much time to execute because the data are divided into N groups. In the following, we describe a method of realizing fast CV in the case of linear regression, which is slightly complicated mathematically but is not limited to LOOCV.

In the following, we consider k nonoverlapping subsets of $\{1, \ldots, N\}$. In CV, one of the subsets is used as the test data, and the other $k - 1$ are training data.

It is known that in linear regression, we have the formula to evaluate the squared loss under cross-validation without actually executing k cycles of evaluations.[1] In the following, by $\|a\|^2$, we denote the squared sum of the N elements in $a \in \mathbb{R}^N$.

Proposition 14 (J. Shao, 1993) *If we divide $\{1, \ldots, N\}$ into nonoverlapping subsets, the sum of the squared loss values evaluated by cross-validation is[2]*

$$\sum_S \|(I - H_S)^{-1} e_S\|^2 ,$$

where for each subset S, $H_S := X_S(X^T X)^{-1} X_S^T$ is the matrix $H = X(X^T X)^{-1} X^T$ that consists of the rows and columns in S, and e_S is the vector $e = y - X\hat{\beta}$ that consists of the rows in S.

In the following, based on the knowledge we have obtained thus far, we show why the proposition holds. We write the matrices that consist of the rows in S and not in $i \notin S$ of $X \in \mathbb{R}^{N \times (p+1)}$ as $X_S \in \mathbb{R}^{r \times (p+1)}$ and $X_{-S} \in \mathbb{R}^{(N-r) \times (p+1)}$, respectively, where r is the cardinality of S. Similarly, we define y_S and y_{-S} for $y \in \mathbb{R}^N$. Then, we have

$$X^T X = \sum_{j=1}^N x_j^T x_j = \sum_{j \in S} x_j^T x_j + \sum_{j \notin S} x_j^T x_j = X_S^T X_S + X_{-S}^T X_{-S} \qquad (4.1)$$

and

$$X^T y = \sum_{j=1}^N x_j^T y_j = \sum_{j \in S} x_j^T y_j + \sum_{j \notin S} x_j^T y_j = X_S^T y_S + X_{-S}^T y_{-S} , \qquad (4.2)$$

where $x_j \in \mathbb{R}^{p+1}$ is a row vector.

Then, we note the following equation:

Proposition 15 (Sherman–Morrison–Woodbury) *Let $m, n \geq 1$. For $A \in \mathbb{R}^{n \times n}$, $U \in \mathbb{R}^{n \times m}$, $C \in \mathbb{R}^{m \times m}$, and $V \in \mathbb{R}^{m \times n}$, we have*

$$(A + UCV)^{-1} = A^{-1} - A^{-1}U(C^{-1} + VA^{-1}U)^{-1}VA^{-1} . \qquad (4.3)$$

For the proof, see the Appendix at the end of this chapter.

---

[1] Many books mention a restrictive formula valid only for LOOCV ($k = N$). This book addresses the general formula applicable to any k.

[2] Linear Model Selection by Cross-Validation Jun Shao, Journal of the American Statistical Association Vol. 88, No. 422 (Jun., 1993), pp. 486–494.

If we apply $n = p + 1, m = r, A = X^T X, C = I, U = X_S^T$, and $V = -X_S$ to (4.3), we have

$$(X_{-S}^T X_{-S})^{-1} = (X^T X)^{-1} + (X^T X)^{-1} X_S^T (I - H_S)^{-1} X_S (X^T X)^{-1} . \qquad (4.4)$$

The following statement assures the nonsingularity of $I - H_S$.

Proposition 16 *Suppose that $X^T X$ is a nonsingular matrix. Then, for each $S \subset \{1, \ldots, N\}$, if $X_{-S}^T X_{-S}$ is invertible, so is $I - H_S$.*

For the proof, see the Appendix at the end of this chapter.

Thus, from (4.1), (4.2), and (4.4), the estimate $\hat{\beta}_{-S} := (X_{-S}^T X_{-S})^{-1} X_{-S}^T y_{-S}$ obtained without using the data in S is as follows:

$$
\begin{aligned}
\hat{\beta}_{-S} &= (X_{-S}^T X_{-S})^{-1} (X^T y - X_S^T y_S) \\
&= \{(X^T X)^{-1} + (X^T X)^{-1} X_S^T (I - H_S)^{-1} X_S (X^T X)^{-1}\}(X^T y - X_S^T y_S) \\
&= \hat{\beta} - (X^T X)^{-1} X_S^T y_S + (X^T X)^{-1} X_S^T (I - H_S)^{-1} (X_S \hat{\beta} - H_S y_S) \\
&= \hat{\beta} - (X^T X)^{-1} X_S^T (I - H_S)^{-1} \{(I - H_S) y_S - X_S \hat{\beta} + H_S y_S\} \\
&= \hat{\beta} - (X^T X)^{-1} X_S^T (I - H_S)^{-1} e_S ,
\end{aligned}
$$

where $\hat{\beta} = (X^T X)^{-1} X^T y$ is the estimate of β obtained using all data, and $e_S := y_S - X_S \hat{\beta}$ is the loss of the data in S when we use the estimate $\hat{\beta}$.

We predict the data that belong to S based on the estimates $\hat{\beta}_{-S}$, and evaluate the residue $y_S - X_S \hat{\beta}_{-S}$:

$$
\begin{aligned}
y_S - X_S \hat{\beta}_{-S} &= y_S - X_S \{\hat{\beta} - (X^T X)^{-1} X_S^T (I - H_S)^{-1} e_S\} \\
&= y_S - X_S \hat{\beta} + X_S (X^T X)^{-1} X_S^T (I - H_S)^{-1} e_S \\
&= e_S + H_S (I - H_S)^{-1} e_S = (I - H_S)^{-1} e_S
\end{aligned}
$$

and its squared sum. Thus, while the residue $y_S - X_S \hat{\beta}$ based on $\hat{\beta}$ obtained using all the data is e_S, the residue $y_S - X_S \hat{\beta}_{-S}$ based on $\hat{\beta}_{-S}$ obtained using the data excluding S is $(I - H_S)^{-1} e_S$, which means that e_S is multiplied by $(I - H_S)^{-1}$, which completes the proof of Proposition 14.

When we obtain the prediction error, we compute $H = X(X^T X)^{-1} X^T$ and $e = (I - H)y$ in advance. We can obtain $(1 - H_S)^{-1} e_S$ for each S by removing some rows of e and some rows and columns of H, and can obtain the squared sum of them over all S.

We construct the following efficient procedure based on Proposition 14:

```
def cv_fast(X,y,k):
    n=len(y)
    m=n/k
    H=X@np.linalg.inv(X.T@X)@X.T
```

```
I=np.diag(np.repeat(1,n))
e=(I-H)@y
I=np.diag(np.repeat(1,m))
S=0
for j in range(k):
    test=np.arange(j*m,(j+1)*m,1,dtype=int)
    S=S+(np.linalg.inv(I-H[test,:][:,test])@e[test]).T@np.linalg.inv(I-H
        [test,test])@e[test]
return S/n
```

```
cv_fast(x,y,10)
```

```
0.04851318320309918
```

Example 42 For each k, we measured how much the execution time differs between the functions cv_fast and cv_linear .

```
# data generation
n=1000; p=5
beta=randn(p+1)
x=randn(n,p)
X=np.insert(x,0,1,axis=1)
y=X@beta+randn(n)
```

```
import time
```

```
U_l=[]; V_l=[]; U_f=[]; V_f=[]
for k in range(2,n+1,1):
    if n%k==0:
        t1=time.time() # Time before processing
        cv_linear(X,y,k)
        t2=time.time() # Time after processing
        U_l.append(k); V_l.append(t2-t1)
        t1=time.time()
        cv_fast(X,y,k)
        t2=time.time()
        U_f.append(k); V_f.append(t2-t1)
plt.plot(U_l,V_l,c="red",label="cv_linear")
plt.plot(U_f,V_f,c="blue",label="cv_fast")
plt.legend()
plt.xlabel("k")
plt.ylabel("execution_time")
plt.title("comparing_between_cv_fast_and_cv_linear")
```

```
Text(0.5, 1.0, 'comparing between cv_fast and cv_linear')
```

The results are shown in Fig. 4.4. There is a large difference at $N = 1000$. The LOOCV of $k = N$ takes the longest processing time, and the difference is large there. However, this difference is not so significant at $N = 100$ because the execution time itself is short. Additionally, since the function cv_fast is specialized for linear regression only, in other problems, a general CV procedure rather than cv_fast will be required.

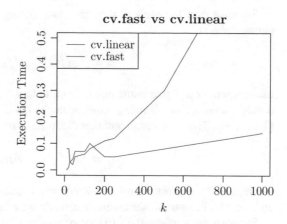

4.3 Bootstrapping

Section 2.7 shows the confidence intervals of the true parameter β and the response
y_* for each covariate x_* in linear regression can be calculated from the observed
data. However, it is not an exaggeration to say that it is rare to be able to do such
things in general settings. To address this issue, we will outline the bootstrap method
and its importance (Fig. 4.5).

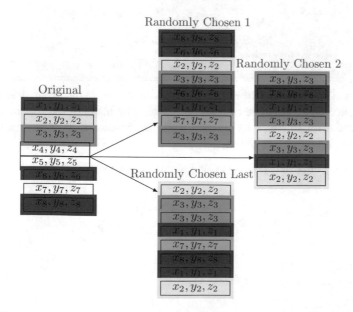

Fig. 4.5 Bootstrapping Multiple data frames of the same size are generated randomly and the
dispersion of the estimated values is observed. Each piece of data in the newly created data frame
must be included in the original data frame

Suppose that for each row in the data frame, we can obtain an estimate of parameter α

$$\hat{\alpha} = f(df_1, \cdots, df_N)$$

using a function f that estimates α from data $df_1, \cdots, df_N$. We consider N randomly chosen rows, allowing duplication, to obtain $(df_{i_1}, \ldots, df_{i_N})$, $i_1, \ldots, i_N \in \{1, \ldots, N\}$. Then, we obtain another estimate of α:

$$\hat{\alpha}_1 = f(df_{i_1}, \ldots, df_{i_N}).$$

We repeat the process r times to obtain the sequence of the estimates $\hat{\alpha}_1, \cdots, \hat{\alpha}_r$ and can obtain an unbiased estimate of variance of the estimates obtained by f.

Suppose we estimate the variance $\sigma^2(\hat{\alpha})$ by

$$\hat{\sigma}^2(\hat{\alpha}) := \frac{1}{r-1} \sum_{h=1}^{r} \left\{ \hat{\alpha}_h - \frac{1}{r} \sum_{f=1}^{r} \hat{\alpha}_f \right\}^2,$$

the following procedure realizes the notion of bootstrap given a function f to estimate, and it evaluates the performance of function f.

```
def bt(df,f,r):
    m=df.shape[0]
    org=f(df,np.arange(0,m,1))
    u=[]
    for j in range(r):
        index=np.random.choice(m,m,replace=True)
        u.append(f(df,index))
    return {'original':org,'bias':np.mean(u)-org,'stderr':np.std(u, ddof=1)}
```

The function `bt` returns the estimate `org` estimated from the original data frame, the difference `bias` between `org` and the arithmetic mean of the r estimates, and the standard error `stderr` of the estimates. The values of the bias and standard error depend on the choice of function f.

Example 43 For N data points $(x_1, y_1), \ldots, (x_N, y_N)$ w.r.t. variables X and Y, we estimate[3]

$$\alpha := \frac{V(Y) - V(X)}{V(X) + V(Y) - 2\mathrm{Cov}(X, Y)},$$

---

[3] In a portfolio, for two brands X and Y, the quantity of X and Y is often estimated.

where $V(\cdot)$ and $\mathrm{Cov}(\cdot, \cdot)$ are the variance and covariance of the variables. Suppose that

$$v_x^2 := \frac{1}{N-1}\left[\sum_{i=1}^{N} x_i^2 - \frac{1}{N}\left\{\sum_{i=1}^{N} x_i\right\}^2\right]$$

$$v_y^2 := \frac{1}{N-1}\left[\sum_{i=1}^{N} y_i^2 - \frac{1}{N}\left\{\sum_{i=1}^{N} y_i\right\}^2\right]$$

$$c_{xy} := \frac{1}{N-1}\left[\sum_{i=1}^{N} x_i y_i - \frac{1}{N}\left\{\sum_{i=1}^{N} x_i\right\}\left\{\sum_{i=1}^{N} y_i\right\}\right],$$

we estimate α by

$$\hat{\alpha} := \frac{v_y^2 - v_x^2}{v_x^2 + v_y^2 - 2c_{xy}}$$

Then, we evaluate the variance via bootstrapping to examine how reliable the estimate $\hat{\alpha}$ is (i.e., how close it is to α).

```
Portfolio=np.loadtxt("Portfolio.csv",delimiter=",",skiprows=1)
def func_1(data,index):
    X=data[index,0]; Y=data[index,1]
    return (np.var(Y, ddof=1)-np.var(X, ddof=1))/(np.var(X, ddof=1)+np.var(Y,
        ddof=1)-2*np.cov(X,Y)[0,1])
bt(Portfolio,func_1,1000)
```

```
{'original': 0.15330230333295436,
 'bias': 0.0063149270038345695,
 'stderr': 0.17757037146622828}
```

If a method for evaluating the estimation error is available, we do not have to assess it by bootstrapping, but for our purposes, let us compare the two to see how correctly bootstrapping performs.

Example 44 We estimated the intercept and slope in the file crime.txt many times by bootstrap, evaluated the dispersion of the estimated values, and compared them with the theoretical values calculated by the sm.OLS function.

In the bootstrap estimation, func_2 estimates the intercept and two slopes when regressing the first variable to the third and fourth variables ($j = 1, 2, 3$) and evaluates its standard deviation.

```
from sklearn import linear_model
```

```
df=np.loadtxt("crime.txt",delimiter="\t")
reg=linear_model.LinearRegression()
```

```
X=df[:,[2,3]]
y=df[:,0]
reg.fit(X,y)
reg.coef_
```

```
array([11.8583308 , -5.97341169])
```

```
for j in range(3):
    def func_2(data,index):
        X=data[index,2:4]; y=data[index,0]
        reg.fit(X,y)
        if j==0:
            return reg.intercept_
        else:
            return reg.coef_[j-1]
    print (bt(df,func_2,1000))
```

```
{'original': 621.4260363802889, 'bias': 39.45710543185794, 'stderr': 220.8724310716836}
{'original': 11.858330796711094, 'bias': -0.4693174397369564, 'stderr': 3.394059052591196}
{'original': -5.973411688164963, 'bias': -0.2157575210725442, 'stderr': 3.166476969985083}
```

```
import statsmodels.api as sm
```

```
n=X.shape[0]
X=np.insert(X,0,1,axis=1)
model=sm.OLS(y,X)
res=model.fit()
print(res.summary())
```

```
                          OLS Regression Results
==============================================================================
Dep. Variable:                      y   R-squared:                       0.325
Model:                            OLS   Adj. R-squared:                  0.296
Method:                 Least Squares   F-statistic:                     11.30
Date:                Mon, 10 Feb 2020   Prob (F-statistic):           9.84e-05
Time:                        00:36:04   Log-Likelihood:                -344.79
No. Observations:                  50   AIC:                             695.6
Df Residuals:                      47   BIC:                             701.3
Df Model:                           2
Covariance Type:            nonrobust
==============================================================================
                 coef    std err          t      P>|t|      [0.025      0.975]
------------------------------------------------------------------------------
const         621.4260    222.685      2.791      0.008     173.441    1069.411
x1             11.8583      2.568      4.618      0.000       6.692      17.024
x2             -5.9734      3.561     -1.677      0.100     -13.138       1.191
==============================================================================
Omnibus:                       14.866   Durbin-Watson:                   1.581
Prob(Omnibus):                  0.001   Jarque-Bera (JB):               16.549
Skew:                           1.202   Prob(JB):                     0.000255
Kurtosis:                       4.470   Cond. No.                         453.
==============================================================================
```

The function `func_2` finds the intercept and the slope of the third and fourth variables at $i = 1, 2, 3$, respectively. In this case, the standard deviation of the intercept and the slopes of the two variables almost match the theoretical values obtained as the output of the `sm.OLS` function. Even if it is a linear regression problem, if the noise does not follow a Gaussian distribution or is not independent, bootstrapping is still useful.

Appendix: Proof of Propositions

Proposition 15 (Sherman–Morrison–Woodbury) *For $m, n \geq 1$ and a matrix $A \in \mathbb{R}^{n \times n}$, $U \in \mathbb{R}^{n \times m}$, $C \in \mathbb{R}^{m \times m}$, $V \in \mathbb{R}^{m \times n}$, we have*

$$(A + UCV)^{-1} = A^{-1} - A^{-1}U(C^{-1} + VA^{-1}U)^{-1}VA^{-1} \tag{4.5}$$

Proof The derivation is due to the following:

$$
\begin{aligned}
&(A + UCV)(A^{-1} - A^{-1}U(C^{-1} + VA^{-1}U)^{-1}VA^{-1}) \\
&= I + UCVA^{-1} - U(C^{-1} + VA^{-1}U)^{-1}VA^{-1} \\
&\quad -UCVA^{-1}U(C^{-1} + VA^{-1}U)^{-1}VA^{-1} \\
&= I + UCVA^{-1} - UC \cdot (C^{-1}) \cdot (C^{-1} + VA^{-1}U)^{-1}VA^{-1} \\
&\quad -UC \cdot VA^{-1}U \cdot (C^{-1} + VA^{-1}U)^{-1}VA^{-1} \\
&= I + UCVA^{-1} - UC(C^{-1} + VA^{-1}U)(C^{-1} + VA^{-1}U)^{-1}VA^{-1} = I.
\end{aligned}
$$

$\square$

Proposition 16 *Suppose that $X^T X$ is a nonsingular matrix. For each $S \subset \{1, \ldots, N\}$, if $X_{-S}^T X_{-S}$ is a nonsingular matrix, so is $I - H_S$.*

Proof For $m, n \geq 1$, $U \in \mathbb{R}^{m \times n}$, and $V \in \mathbb{R}^{n \times m}$, we have

$$
\begin{bmatrix} I & 0 \\ V & I \end{bmatrix} \begin{bmatrix} I + UV & U \\ 0 & I \end{bmatrix} \begin{bmatrix} I & 0 \\ -V & I \end{bmatrix} = \begin{bmatrix} I + UV & U \\ V + VUV & VU + I \end{bmatrix} \begin{bmatrix} I & 0 \\ -V & I \end{bmatrix}
$$

$$
= \begin{bmatrix} I & U \\ 0 & I + VU \end{bmatrix}.
$$

Combined with Proposition 2, we have

$$\det(I + UV) = \det(I + VU). \tag{4.6}$$

Therefore, from Proposition 2, we have

$$\det(X_{-S}^T X_{-S}) = \det(X^T X - X_S^T X_S)$$
$$= \det(X^T X) \det(I - (X^T X)^{-1} X_S^T X_S)$$
$$= \det(X^T X) \det(I - X_S(X^T X)^{-1} X_S^T),$$

where the last transformation is due to (4.6). Hence, from Proposition 1, if $X_{-S}^T X_{-S}$ and $X^T X$ are nonsingular, so is $I - H_S$.   □

Exercises 32–39

32. Let $m, n \geq 1$. Show that for matrix $A \in \mathbb{R}^{n \times n}$, $U \in \mathbb{R}^{n \times m}$, $C \in \mathbb{R}^{m \times m}$, $V \in \mathbb{R}^{m \times n}$,

$$(A + UCV)^{-1} = A^{-1} - A^{-1}U(C^{-1} + VA^{-1}U)^{-1}VA^{-1} \qquad (4.7)$$

(Sherman–Morrison–Woodbury). Hint: Continue the following:

$$(A + UCV)(A^{-1} - A^{-1}U(C^{-1} + VA^{-1}U)^{-1}VA^{-1})$$
$$= I + UCVA^{-1} - U(C^{-1} + VA^{-1}U)^{-1}VA^{-1}$$
$$-UCVA^{-1}U(C^{-1} + VA^{-1}U)^{-1}VA^{-1}$$
$$= I + UCVA^{-1} - UC \cdot (C^{-1}) \cdot (C^{-1} + VA^{-1}U)^{-1}VA^{-1}$$
$$-UC \cdot VA^{-1}U \cdot (C^{-1} + VA^{-1}U)^{-1}VA^{-1}.$$

33. Let S be a subset of $\{1, \ldots, N\}$ and write the matrices $X \in \mathbb{R}^{(N-r) \times (p+1)}$ that consist of the rows in S and the rows not in S as $X_S \in \mathbb{R}^{r \times (p+1)}$ and $X_{-S} \in \mathbb{R}^{(N-r) \times (p+1)}$, respectively, where r is the number of elements in S. Similarly, we divide $y \in \mathbb{R}^N$ into y_S and y_{-S}.

(a) Show

$$(X_{-S}^T X_{-S})^{-1} = (X^T X)^{-1} + (X^T X)^{-1} X_S^T (I - H_S)^{-1} X_S (X^T X)^{-1},$$

where $H_S := X_S(X^T X)^{-1} X_S^T$ is the matrix that consists of the rows and columns in S of $H = X(X^T X)^{-1} X^T$. Hint: Apply $n = p + 1$, $m = r$, $A = X^T X$, $C = I$, $U = X_S^T$, $V = -X_S$ to (4.3).

(b) For $e_S := y_S - \hat{y}_S$ with $\hat{y}_S = X_S \hat{\beta}$, show the equation

$$\hat{\beta}_{-S} = \hat{\beta} - (X^T X)^{-1} X_S^T (I - H_S)^{-1} e_S$$

Hint: From $X^T X = X_S^T X_S + X_{-S}^T X_{-S}$ and $X^T y = X_S^T y_S + X_{-S}^T y_{-S}$,

$$\hat{\beta}_{-S} = \{(X^T X)^{-1} + (X^T X)^{-1} X_S^T (I - H_S)^{-1} X_S (X^T X)^{-1}\}(X^T y - X_S^T y_S)$$

$$= \hat{\beta} - (X^T X)^{-1} X_S^T (I - H_S)^{-1} (X_S \hat{\beta} - H_S y_S)$$

$$= \hat{\beta} - (X^T X)^{-1} X_S^T (I - H_S)^{-1} \{(I - H_S) y_S - X_S \hat{\beta} + H_S y_S\}.$$

34. By showing $y_S - X_S \hat{\beta}_{-S} = (I - H_S)^{-1} e_S$, prove that the squared sum of the groups in CV is $\sum_S \|(I - H_S)^{-1} e_S\|^2$, where $\|a\|^2$ denotes the squared sum of the elements in $a \in \mathbb{R}^N$.

35. Fill in the blanks below and execute the procedure in Problem 34. Observe that the squared sum obtained by the formula and by the general cross-validation method coincide.

```
n=1000; p=5
X=np.insert(randn(n,p),0,1,axis=1)
beta=randn(p+1).reshape(-1,1)
y=X@beta+0.2*randn(n).reshape(-1,1)
y=y[:,0]
```

```
# Conventional CV
def cv_linear(X,y,K):
    n=len(y); m=int(n/K)
    S=0
    for j in range(K):
        test=list(range(j*m,(j+1)*m))  # indices for test data
        train=list(set(range(n))-set(test))     # indices for train data
        beta=np.linalg.inv(X[train,].T@X[train,])@X[train,].T@y[train]
        e=y[test]-X[test,]@beta
        S=S+np.linalg.norm(e)**2
    return S/n
```

```
# Fast CV
def cv_fast(X,y,k):
    n=len(y)
    m=n/k
    H=X@np.linalg.inv(X.T@X)@X.T
    I=np.diag(np.repeat(1,n))
    e=(I-H)@y
    I=np.diag(np.repeat(1,m))
    S=0
    for j in range(k):
        test=np.arange(j*m,(j+1)*m,1,dtype=int)
        S=S+(np.linalg.inv(I-H[test,test])@e[test]).T@np.linalg.inv(I-H[
            test,test])@e[test]
    return S/n
```

Moreover, we wish to compare the speeds of the functions `cv_linear` and `cv_fast`. Fill in the blanks below to complete the procedure and draw the graph.

```
import time
```

```
U_l=[]; V_l=[]
for k in range(2,n+1,1):
    if n%k==0:
        t1=time.time() # time before processing
        cv_linear(X,y,k)
        t2=time.time() #  time after processing
        U_l.append(k); V_l.append(t2-t1)

# some blanks #

plt.plot(U_l,V_l,c="red",label="cv_linear")
plt.legend()
plt.xlabel("k")
plt.ylabel("execution_time")
plt.title("compairing_between_cv_fast_and_cv_linear")
```

```
Text(0.5, 1.0, 'compairing between cv_fast and cv_linear')
```

36. How much the prediction error differs with k in the k-fold CV depends on the data. Fill in the blanks and draw the graph that shows how the CV error changes with k. You may use either the function `cv_linear` or `cv_fast`.

```
n=100; p=5
plt.ylim(0.3,1.5)
plt.xlabel("k")
plt.ylabel("values_of_CV")
for j in range(2,11,1):
    X=randn(n,p)
    X=np.insert(X,0,1,axis=1)
    beta=randn(p+1)
    y=X@beta+randn(n)
    U=[]; V=[]
    for k in range(2,n+1,1):
        if n%k==0:
            # blank #
    plt.plot(U,V)
```

37. We wish to know how the error rate changes with K in the K-nearest neighbor method when 10-fold CV is applied for the Fisher's Iris data set. Fill in the blanks, execute the procedure, and draw the graph.

```
from sklearn.datasets import load_iris
```

```
iris=load_iris()
iris.target_names
x=iris.data
y=iris.target
n=x.shape[0]
```

```
order=np.random.choice(n,n,replace=False) # rearrange
x=x[index,]
y=y[index]
```

```
U=[]
V=[]
top_seq=list(range(0,135,15))
for k in range(1,11,1):
    S=0
    for top in top_seq:
        test=# blank(1) #
        train=list(set(range(150))-set(test))
        knn_ans=knn(x[train,],y[train],x[test,],k=k)
        ans=# blank(2) #
        S=S+np.sum(knn_ans!=ans)
    S=S/n
    U.append(k)
    V.append(S)
plt.plot(U,V)
plt.xlabel("K")
plt.ylabel("error_rate_")
plt.title("Assessment_of_error_rate_by_CV")
```

```
Text(0.5, 1.0, 'Assessment of error rate by CV')
```

38. We wish to estimate the standard deviation of the quantity below w.r.t. X, Y based on N data.

$$\frac{v_y - v_x}{v_x + v_y - 2v_{xy}}, \quad \begin{cases} v_x := \dfrac{1}{N-1}\left[\displaystyle\sum_{i=1}^{N} X_i^2 - \dfrac{1}{N}\left\{\sum_{i=1}^{N} X_i\right\}^2\right] \\[4mm] v_y := \dfrac{1}{N-1}\left[\displaystyle\sum_{i=1}^{N} Y_i^2 - \dfrac{1}{N}\left\{\sum_{i=1}^{N} Y_i\right\}^2\right] \\[4mm] v_{xy} := \dfrac{1}{N-1}\left[\displaystyle\sum_{i=1}^{N} X_i Y_i - \dfrac{1}{N}\left\{\sum_{i=1}^{N} X_i\right\}\left\{\sum_{i=1}^{N} Y_i\right\}\right] \end{cases}$$

To this end, allowing duplication, we randomly choose N data in the data frame r times and estimate the standard deviation (Bootstrap). Fill in the blanks (1)(2) to complete the procedure and observe that it estimates the standard deviation.

```
def bt(df,f,r):
    m=df.shape[0]
    org=# blank(1) #
    u=[]
    for j in range(r):
        index=np.random.choice(# blank(2) #)
        u.append(f(df,index))
    return {'original':org,'bias':np.mean(u)-org,'stderr':np.std(u, ddof
        =1)}
```

```
def func_1(data,index):
    X=data[index,0]; Y=data[index,1]
    return (np.var(Y, ddof=1)-np.var(X, ddof=1))/(np.var(X, ddof=1)+np.
       var(Y, ddof=1)-2*np.cov(X,Y)[0,1])
```

```
Portfolio=np.loadtxt("Portfolio.csv",delimiter=",",skiprows=1)
bt(Portfolio,func_1,1000)
```

39. For linear regression, if we assume that the noise follows a Gaussian distribution, we can compute the theoretical value of the standard deviation. We wish to compare the value with the one obtained by bootstrap. Fill in the blanks and execute the procedure. What are the three kinds of data that appear first?

```
from sklearn import linear_model
```

```
df=np.loadtxt("crime.txt",delimiter="\t")
reg=linear_model.LinearRegression()
X=df[:,[2,3]]
y=df[:,0]
reg.fit(X,y)
reg.coef_
```

```
array([11.8583308 , -5.97341169])
```

```
for j in range(3):
    def func_2(data,index):
        X=data[index,2:4]; y=## blank ##
        reg.fit(X,y)
        if j==0:
            return reg.intercept_
        else:
            return reg.coef_[j-1]
    print(bt(df,func_2,1000))
```

```
import statsmodels.api as sm
```

```
n=X.shape[0]
X=np.insert(X,0,1,axis=1)
model=sm.OLS(y,X)
res=model.fit()
print(res.summary())
```

Chapter 5
Information Criteria

Abstract Until now, from the observed data, we have considered the following cases:

- Build a statistical model and estimate the parameters contained in it.
- Estimate the statistical model.

In this chapter, we consider the latter for linear regression. The act of finding rules from observational data is not limited to data science and statistics. However, many scientific discoveries are born through such processes. For example, the writing of the theory of elliptical orbits, the law of constant area velocity, and the rule of harmony in the theory of planetary motion published by Kepler in 1596 marked the transition from the dominant theory to the planetary motion theory. While the explanation by the planetary motion theory was based on countless theories based on philosophy and thought, Kepler's law solved most of the questions at the time with only three laws. In other words, as long as it is a law of science, it must not only be able to explain phenomena (fitness), but it must also be simple (simplicity). In this chapter, we will learn how to derive and apply the AIC and BIC, which evaluate statistical models of data and balance fitness and simplicity.

5.1 Information Criteria

Information criterion is generally defined as an index for evaluating the validity of a statistical model from observation data. Akaike's information criterion (AIC) and the Bayesian information criterion (BIC) are well known. An information criterion often refers to the evaluation of both how much the statistical model explains the data (fitness) and how simple the statistical model is (simplicity). AIC and BIC are standard except for the difference in how they are balanced.

The same can be done with the cross-validation approach discussed in Chap. 4, which is superior to the information criteria in versatility but does not explicitly control the balance between fitness and simplicity.

One of the most important problems in linear regression is to select some p covariates based on N observations $(x_1, y_1), \ldots, (x_N, y_N) \in \mathbb{R}^p \times \mathbb{R}$. The reason

© The Author(s), under exclusive license to Springer Nature Singapore Pte Ltd. 2021
J. Suzuki, *Statistical Learning with Math and Python*,
https://doi.org/10.1007/978-981-15-7877-9_5

why there should not be too many covariates is that they overfit the data and try to explain the noise fluctuation by other covariates. Thus, we need to recognize the exact subset.

However, it is not easy to choose $S \subseteq \{1, \ldots, p\}$ from the 2^p subsets

$$\{\}, \{1\}, \ldots, \{p\}, \{1, 2\}, \ldots, \{1, \ldots, p\}$$

when p is large because 2^p increases exponentially with p. We express the fitness and simplicity by the RSS value $RSS(S)$ based on the subset S and the cardinality[1] $k(S) := |S|$ of S. Then, we have that

$$S \subseteq S' \implies \begin{cases} RSS(S) \geq RSS(S') \\ k(S) \leq k(S'), \end{cases}$$

which means that the larger the $k = k(S)$, the smaller $\hat{\sigma}_k^2 = \dfrac{RSS_k}{N}$ is, where $RSS_k := \min_{k(S)=k} RSS(S)$. The AIC and BIC are defined by

$$AIC := N \log \hat{\sigma}_k^2 + 2k \tag{5.1}$$

$$BIC := N \log \hat{\sigma}_k^2 + k \log N , \tag{5.2}$$

and the coefficient of determination

$$1 - \frac{RSS_k}{TSS}$$

increases monotonically with k and reaches its maximum value at $k = p$. However, the AIC and BIC values decrease before reaching the minimum at some $0 \leq k \leq p$ and increase beyond that point, where the k values that minimize the AIC and BIC are minimized are generally different. The adjusted coefficient of determination maximizes

$$1 - \frac{RSS_k/(N - k - 1)}{TSS/(N - 1)}$$

at some $0 \leq k \leq p$, which is often much larger than those of the AIC and BIC.

Example 45 The following data fields are from the Boston dataset in the Python `sklearn` package. We assume that the first thirteen variables and the last variable are covariates and a response, respectively.

We construct the following procedure to find the set of covariates that minimizes the AIC. In particular, we execute `itertools.combinations(range(p),k)`

---

[1]By $|S|$, we mean the cardinality of set S.

Column #	Variable	Meaning of the variable
1	CRIM	Per capita crime rate by town
2	ZN	Proportion of residential land zoned for lots over 25,000 sq. ft.
3	INDUS	Proportion of nonretail business acres per town
4	CHAS	Charles River dummy variable (1 if the tract bounds the river; 0 otherwise)
5	NOX	Nitric oxide concentration (parts per 10 million)
6	RM	Average number of rooms per dwelling
7	AGE	Proportion of owner-occupied units built prior to 1940
8	DIS	Weighted distances to five Boston employment centers
9	RAD	Index of accessibility to radial highways
10	TAX	Full-value property tax rate per $10,000
11	PTRATIO	Student–teacher ratio by town
12	B	$1000(Bk - 0.63)^2$, where Bk is the proportion of black people by town
13	LSTAT	% lower status of the population
14	MEDV	Median value of owner-occupied homes in $1000s

to obtain a matrix of size $k \times \binom{p}{k}$ that has subsets $\{1, \cdots, p\}$ of size k in its columns to find the minimum value $\hat{\sigma}_k^2$ over S such that $|S| = k$.

```
from sklearn.linear_model import LinearRegression
import itertools   # enumerate combinations
```

```
res=LinearRegression()
```

```
def RSS_min(X,y,T):
    S_min=np.inf
    m=len(T)
    for j in range(m):
        q=T[j]
        res.fit(X[:,q],y)
        y_hat=res.predict(X[:,q])
        S=np.linalg.norm(y_hat-y)**2
        if S<S_min:
            S_min=S
            set_q=q
    return(S_min,set_q)
```

We compute $N \log \hat{\sigma}_k^2 + 2k$ for S and find the value of k that minimizes it among $k = 0, 1, \ldots, p$.

```
from sklearn.datasets import load_boston
```

```
boston=load_boston()
X=boston.data[:,[0,2,4,5,6,7,9,10,11,12]]
y=boston.target
```

```
n,p=X.shape
AIC_min=np.inf
for k in range(1,p+1,1):
    T=list(itertools.combinations(range(p),k))
    # each column has combinations (k from p)
    S_min,set_q=RSS_min(X,y,T)
    AIC=n*np.log(S_min/n)+2*k   ##
    if AIC<AIC_min:
        AIC_min=AIC
        set_min=set_q
print(AIC_min,set_min)
```

```
4770.415163216072 (0, 2, 3, 5, 7, 8, 9)
```

If we replace the line `n*np.log(S.min)+2*k` marked by `##` with `n*np.log(S.min)+k*np.log(N)`, then the quantity becomes the BIC. To maximize the adjusted coefficient of determination, we may update it as follows:

```
y_bar=np.mean(y)
TSS=np.linalg.norm(y-y_bar)**2
D_max=-np.inf
for k in range(1,p+1,1):
    T=list(itertools.combinations(range(p),k))
    S_min,set_q=RSS_min(X,y,T)
    D=1-(S_min/(n-k-1))/(TSS/(n-1))
    if D>D_max:
        D_max=D
        set_max=set_q
print(D_max,set_max)
```

```
0.9999988717090253 (0, 1, 2, 3, 4, 5, 6, 7, 8, 9)
```

For each $k = 0, 1, \ldots, p$, we find the S that minimizes $RSS(S)$ such that $|S| = k$ for each $k = 0, 1, \ldots, p$ and compute the adjusted coefficient of determination from RSS_k. Then, we obtain the maximum value of the adjusted coefficients of determination over $k = 0, 1, \ldots, p$.

On the other hand, the BIC (5.2) is used as often as the AIC (5.1). The difference is only in the balance between fitness $\log \hat{\sigma}_k^2$ and simplicity k.

We see that the schools with 200 and 100 points for English and math, respectively, on the entrance examination choose different applicants from the schools with 100 and 200 points for English and math, respectively. Similarly, the statistical models selected by the AIC and BIC are different. Since the BIC has a more significant penalty for the simplicity k, the selected k is smaller, and the chosen model is simpler than that chosen by the AIC.

More importantly, the BIC converges to the correct model when the number of samples N is large (consistency), but the AIC does not. The AIC was developed

Fig. 5.1 The BIC is larger than the AIC, but the BIC chooses a simpler model with fewer variables than the AIC

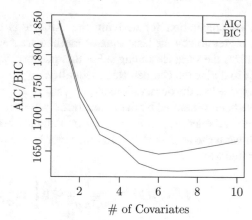

Changes of AIC/BIC with # of Covariates

to minimize the prediction error (Sect. 5.4). Even if the statistical model selected is incorrect, the squared error in the test data may be small for the finite number of samples N, which is an advantage of the AIC. It is essential to use the specific information criteria according to their intended purposes, and it is meaningless to discuss which one is superior to the other (Fig. 5.1).

Example 46

```
def IC(X,y,k):
    n,p=X.shape
    T=list(itertools.combinations(range(p),k))
    S,set_q=RSS_min(X,y,T)
    AIC=n*np.log(S/n)+2*k
    BIC=n*np.log(S/n)+k*np.log(n)
    return {'AIC':AIC,'BIC':BIC}
```

```
AIC_seq=[]; BIC_seq=[]
for k in range(1,p+1,1):
    AIC_seq.append(IC(X,y,k)['AIC'])
    BIC_seq.append(IC(X,y,k)['BIC'])
x_seq=np.arange(1,p+1,1)
plt.plot(x_seq,AIC_seq,c="red",label="AIC")
plt.plot(x_seq,BIC_seq,c="blue",label="BIC")
plt.xlabel("_the_number_of_variables_")
plt.ylabel("values_of_AIC/BIC")
plt.title("changes_of_the_number_of_variables_and_AIC_and_BIC")
plt.legend()
```

5.2   Efficient Estimation and the Fisher Information Matrix

Next, as preparation for deriving the AIC, it is deduced that the estimates of linear regression by the least squares method are a so-called efficient estimator that minimizes the variance among s. For this purpose, we define the Fisher information matrix to derive the Cramér–Rao inequality.

Suppose that the observations $x_1, \ldots, x_N \in \mathbb{R}^{p+1}$ (row vector) and $y_1, \ldots, y_N \in \mathbb{R}$ have been generated by the realizations $y_i = x_i \beta + \beta_0 + e_i, i = 1, \ldots, N$ with random variables $e_1, \ldots, e_N \sim N(0, \sigma^2)$ and unknown constants $\beta_0 \in \mathbb{R}$, $\beta \in \mathbb{R}^p$, which we write as $\beta \in \mathbb{R}^{p+1}$. In other words, the probability density function can be written as

$$f(y_i | x_i, \beta) := \frac{1}{\sqrt{2\pi\sigma^2}} \exp\left\{-\frac{1}{2\sigma^2} \|y_i - x_i \beta\|^2\right\}$$

$$X = \begin{bmatrix} x_1 \\ \vdots \\ x_N \end{bmatrix} \in \mathbb{R}^{N \times (p+1)}, \quad y = \begin{bmatrix} y_1 \\ \vdots \\ y_N \end{bmatrix} \in \mathbb{R}^N, \quad \beta = \begin{bmatrix} \beta_0 \\ \beta_1 \\ \vdots \\ \beta_p \end{bmatrix} \in \mathbb{R}^{p+1}.$$

In the least squares method, we estimated β by $\hat{\beta} = (X^T X)^{-1} X^T y$ if $X^T X$ is nonsingular (Proposition 11).

We claim that $\hat{\beta}$ coincides with the $\beta \in \mathbb{R}^{p+1}$ that maximizes the likelihood

$$L := \prod_{i=1}^{N} f(y_i | x_i, \beta).$$

In fact, the log-likelihood is written by

$$l := \log L = -\frac{N}{2} \log(2\pi\sigma^2) - \frac{1}{2\sigma^2} \|y - X\beta\|^2 .$$

If $\sigma^2 > 0$ is fixed, maximizing this value is equivalent to minimizing $\|y - X\beta\|^2$. Moreover, if we partially differentiate l w.r.t. σ^2, we have that

$$\frac{\partial l^2}{\partial \sigma^2} = -\frac{N}{2\sigma^2} + \frac{\|y - X\beta\|^2}{2(\sigma^2)^2} = 0 .$$

Thus, using $\hat{\beta} = (X^T X)^{-1} X^T y$, we find that

$$\hat{\sigma}^2 := \frac{1}{N} \|y - X\hat{\beta}\|^2 = \frac{RSS}{N}$$

is the maximum likelihood estimate of σ^2. In Chap. 2, we derived $\hat{\beta} \sim N(\beta, \sigma^2(X^T X)^{-1})$, which means that $\hat{\beta}$ is an unbiased estimator and the covariance matrix is $\sigma^2(X^T X)^{-1}$. In general, if the variance is minimized among the unbiased estimators, it is called an efficient estimator. In the following, we show that the estimate $\hat{\beta}$ is an efficient estimator.

Let ∇l be the vector consisting of $\frac{\partial l}{\partial \beta_j}$, $j = 0, 1, \ldots, p$. We refer to the covariance matrix J of ∇l divided by N as the Fisher information matrix. For $f^N(y|x, \beta) := \prod_{i=1}^{N} f(y_i|x_i, \beta)$, we have

$$\nabla l = \frac{\nabla f^N(y|x, \beta)}{f^N(y|x, \beta)}.$$

Suppose that the order between the derivative w.r.t. β and the integral w.r.t. y can be switched.[2] If we partially differentiate both sides of $\int f^N(y|x, \beta)dy = 1$ w.r.t. β, we have that $\int \nabla f^N(y|x, \beta)dy = 0$. On the other hand, we have that

$$E\nabla l = \int \frac{\nabla f^N(y|x, \beta)}{f^N(y|x, \beta)} f^N(y|x, \beta)dy = \int \nabla f^N(y|x, \beta)dy = 0 \qquad (5.3)$$

and

$$0 = \nabla \otimes [E\nabla l] = \nabla \otimes \int (\nabla l) f^N(y|x, \beta)dy$$

$$= \int (\nabla^2 l) f^N(y|x, \beta)dy + \int (\nabla l)\{\nabla f^N(y|x, \beta)\}dy$$

$$= E[\nabla^2 l] + E[(\nabla l)^2]. \qquad (5.4)$$

In particular, (5.4) implies that

$$J = \frac{1}{N} E[(\nabla l)^2] = -\frac{1}{N} E[\nabla^2 l]. \qquad (5.5)$$

Example 47 For linear regression, we analyze (5.5):

$$\nabla l = \frac{1}{\sigma^2} \sum_{i=1}^{N} x_i^T (y_i - x_i \beta)$$

---

[2]In many practical situations, including linear regression, no problem occurs.

$$\nabla^2 l = -\frac{1}{\sigma^2} \sum_{i=1}^{N} x_i^T x_i = -\frac{1}{\sigma^2} X^T X$$

$$E[\nabla l] = \frac{1}{\sigma^2} \sum_{i=1}^{N} x_i^T E(y_i - x_i \beta) = 0$$

$$E[(\nabla l)(\nabla l)^T] = \frac{1}{(\sigma^2)^2} E\left[\sum_{i=1}^{N} x_i^T (y_i - x_i \beta) \{ \sum_{j=1}^{N} x_j^T (y_j - x_i \beta) \}^T \right]$$

$$= \frac{1}{(\sigma^2)^2} \sum_{i=1}^{N} x_i^T E(y_i - x_i \beta)(y_i - x_i \beta)^T x_i = \frac{1}{(\sigma^2)^2} \sum_{i=1}^{N} x_i^T \sigma^2 I x_i$$

$$= \frac{1}{\sigma^2} \sum_{i=1}^{N} x_i^T x_i = \frac{1}{\sigma^2} X^T X$$

$$V[\nabla l] = \frac{1}{(\sigma^2)^2} \sum_{i=1}^{N} x_i^T E(y_i - x_i \beta)(y_i - x_i \beta)^T x_i$$

$$= \frac{1}{(\sigma^2)^2} \sum_{i=1}^{N} x_i^T \sigma^2 I x_i = \frac{1}{\sigma^2} X^T X.$$

In general, we have the following statement.

Proposition 17 (Cramér–Rao Inequality) *Any covariance matrix* $V(\tilde{\beta})$ $\in$ $\mathbb{R}^{(p+1)\times(p+1)}$ *w.r.t. an unbiased estimate is not below the inverse of the Fisher information matrix:*

$$V(\tilde{\beta}) \geq (NJ)^{-1},$$

where an inequality between matrices ≥ 0 *implies that the difference is nonnegative definite.*

Note that the least squares estimate satisfies the equality part of the inequality. To this end, if we partially differentiate both sides of

$$\int \tilde{\beta}_i f^N(y|x, \beta) dy = \beta_i$$

w.r.t. β_j, we have the following equation:

$$\int \tilde{\beta}_i \frac{\partial}{\partial \beta_j} f^N(y|x, \beta) dy = \begin{cases} 1, & i = j \\ 0, & i \neq j. \end{cases}$$

If we write this equation in terms of its covariance matrix, we have that $E[\tilde{\beta}(\nabla l)^T] = I$, where I is a unit matrix of size $(p+1)$. Moreover, from $E[\nabla l] = 0$ (5.3), we rewrite the above equation as

$$E[(\tilde{\beta} - \beta)(\nabla l)^T] = I .\qquad(5.6)$$

Then, the covariance matrix of the vector of size $2(p+1)$ that consists of $\tilde{\beta} - \beta$ and ∇l is

$$\begin{bmatrix} V(\tilde{\beta}) & I \\ I & NJ \end{bmatrix} .$$

Note that because both $V(\tilde{\beta})$ and J are covariance matrices, they are nonnegative definite. Finally, we claim that both sides of

$$\begin{bmatrix} V(\tilde{\beta}) - (NJ)^{-1} & 0 \\ 0 & NJ \end{bmatrix} = \begin{bmatrix} I & -(NJ)^{-1} \\ 0 & I \end{bmatrix} \begin{bmatrix} V(\tilde{\beta}) & I \\ I & NJ \end{bmatrix} \begin{bmatrix} I & 0 \\ -(NJ)^{-1} & I \end{bmatrix}$$

are nonnegative definite. In fact, for an arbitrary $x \in \mathbb{R}^n$, if $x^T A x \geq 0$, for an arbitrary $B \in \mathbb{R}^{n \times m}$, and $y \in \mathbb{R}^m$, we have that $y^T B^T A B y \geq 0$, which means that $V(\tilde{\beta}) - (NJ)^{-1}$ is nonnegative definite (for $x, y \in \mathbb{R}^{p+1}$, the inequality $x^T\{V(\tilde{\beta}) - (NJ)^{-1}\}x + y^T NJy \geq 0$ should hold even if $y = 0$). This completes the proof of Proposition 17.

5.3 Kullback–Leibler Divergence

For probability density functions $f, g \in \mathbb{R}$, we refer to

$$D(f \| g) := \int_{-\infty}^{\infty} f(x) \log \frac{f(x)}{g(x)} dx$$

as to the Kullback–Leibler (KL) divergence, which is defined if the following condition is met:

$$\int_S f(x)dx > 0 \Longrightarrow \int_S g(x)dx > 0$$

for an arbitrary $S \subseteq \mathbb{R}$.

In general, since $D(f \| g)$ and $D(g \| f)$ do not coincide, the KL divergence is not a distance. However, $D(f \| g) \geq 0$ and is equal to zero if and only if f and g

Fig. 5.2 $y = x - 1$ is beyond
$y = \log x$ except $x = 1$

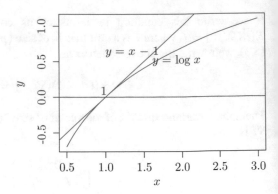

coincide. In fact, we have from $\log x \leq x - 1$, $x > 0$ (Fig. 5.2) that

$$\int_{-\infty}^{\infty} f(x) \log \frac{f(x)}{g(x)} dx = -\int_{-\infty}^{\infty} f(x) \log \frac{g(x)}{f(x)} dx \geq -\int_{-\infty}^{\infty} f(x) \left(\frac{g(x)}{f(x)} - 1 \right) dx$$

$$= -\int_{-\infty}^{\infty} (g(x) - f(x)) dx = 1 - 1 = 0.$$

In the following, we compute the KL divergence value $D(\beta \| \gamma)$ of a parameter γ when the true parameter is β.

Proposition 18 *For covariates $x_1, \ldots, x_N$, if the responses are $z_1, \ldots, z_N$, the likelihood $-\sum_{i=1}^{N} \log f(z_i | x_i, \gamma)$ of $\gamma \in \mathbb{R}^{p+1}$ is*

$$\frac{N}{2} \log 2\pi \sigma^2 + \frac{1}{2\sigma^2} \|z - X\beta\|^2 - \frac{1}{\sigma^2} (\gamma - \beta)^T X^T (z - X\beta) + \frac{1}{2\sigma^2} (\gamma - \beta)^T X^T X (\gamma - \beta) \tag{5.7}$$

for an arbitrary $\beta \in \mathbb{R}^{p+1}$.

For the proof, see the Appendix at the end of this chapter.

We assume that $z_1, \ldots, z_N$ has been generated by $f(z_1 | x_1, \beta), \ldots, f(z_N | x_N, \beta)$, where the true parameter is β. Then, the average of (5.7) w.r.t. $Z_1 = z_1, \ldots, Z_N = z_N$ is

$$- E_Z \sum_{i=1}^{N} \log f(Z_i | x_i, \gamma) = \frac{N}{2} \log(2\pi\sigma^2 e) + \frac{1}{2\sigma^2} \|X(\gamma - \beta)\|^2 \tag{5.8}$$

since the averages of $z - X\beta$ and $\|z - X\beta\|^2$ are 0 and $N\sigma^2$, respectively.

Moreover, the value of (5.8) can be written as

$$-\sum_{i=1}^{N} \int_{-\infty}^{\infty} f(z|x_i, \beta) \log f(z|x_i, \gamma) dz .$$

Since $\sum_{i=1}^{N} \int_{-\infty}^{\infty} f(z|x_i, \beta) \log f(z|x_i, \beta) dz$ is a constant that does not depend on γ, we only need to choose a parameter γ so that the sum of Kullback–Leibler divergence

$$E_Z \sum_{i=1}^{N} \log \frac{f(z|x_i, \beta)}{f(z|x_i, \gamma)} = \sum_{i=1}^{N} \int_{-\infty}^{\infty} f(z|x_i, \beta) \log \frac{f(z|x_i, \beta)}{f(z|x_i, \gamma)} dz = \frac{1}{2\sigma^2} \|X(\gamma - \beta)\|^2$$

(5.9)

is minimized.

5.4 Derivation of Akaike's Information Criterion

In general, the true parameters β are unknown and should be estimated. In the following, our goal is to choose a γ among the s so that (5.9) is minimized on average.

In general, for random variables $U, V \in \mathbb{R}^N$, we have that

$$\{E[U^T V]\}^2 \leq E[\|U\|^2]E[\|V\|^2] \qquad \text{(Schwarz's inequality)} .$$

In fact, in the quadratic equation w.r.t. t

$$E(tU + V)^2 = t^2 E[\|U\|^2] + 2t E[U^T V] + E[\|V\|^2] = 0,$$

at most one solution exists, so the determinant is not positive. If we let $U = X(X^T X)^{-1} \nabla l$ and $V = X(\tilde{\beta} - \beta)$, then we have

$$\{E[(\tilde{\beta} - \beta)^T \nabla l]\}^2 \leq E\|X(X^T X)^{-1} \nabla l\|^2 E\|X(\tilde{\beta} - \beta)\|^2 .$$

(5.10)

In the following, we use the fact that for matrices $A = (a_{i,j})$ and $B = (b_{i,j})$, if the products AB and BA are defined, then both traces are $\sum_i \sum_j a_{i,j} b_{j,i}$ and coincide. Now, the traces of the left-hand and right-hand sides of (5.6) are

$$\text{trace}\{E[(\tilde{\beta} - \beta)(\nabla l)^T]\} = \text{trace}\{E[(\nabla l)^T (\tilde{\beta} - \beta)]\} = E[(\tilde{\beta} - \beta)^T (\nabla l)]$$

and $p + 1$, which means that

$$E[(\tilde{\beta} - \beta)^T (\nabla l)] = p + 1 .\tag{5.11}$$

Moreover, we have that

$$E\|X(X^T X)^{-1}\nabla l\|^2 = E\left\{\text{trace}\left[(\nabla l)^T (X^T X)^{-1} X^T X (X^T X)^{-1}\nabla l\right]\right\}$$
$$= \text{trace}\{(X^T X)^{-1} E(\nabla l)(\nabla l)^T\}$$
$$= \text{trace}\{(X^T X)^{-1}\sigma^{-2} X^T X\} = \text{trace}\{\sigma^{-2} I\} = (p + 1)/\sigma^2 .$$
$$\tag{5.12}$$

Thus, from (5.10), (5.11), and (5.12), we have that

$$E\{\|X(\tilde{\beta} - \beta)\|^2\} \geq (p + 1)\sigma^2$$

On the other hand, if we apply the least squares method: $\hat{\beta} = (X^T X)^{-1} X^T y$, we have that

$$E\|X(\hat{\beta} - \beta)\|^2 = E[\text{trace}\,(\hat{\beta} - \beta)^T X^T X (\hat{\beta} - \beta)]$$
$$= \text{trace}\,(V[\hat{\beta}]X^T X) = \text{trace}\,(\sigma^2 I) = (p + 1)\sigma^2,$$

and the equality holds.

The goal of Akaike's information criterion is to minimize the quantity

$$\frac{N}{2}\log 2\pi\sigma^2 + \frac{1}{2}(p + 1)$$

obtained by replacing the second term of (5.8) with its average. In particular, for the problem of variable selection, the number of the covariates is not p, but any $0 \leq k \leq p$. Hence, we choose the k that minimizes

$$N\log\sigma_k^2 + k .\tag{5.13}$$

Note that the value of $\sigma_k^2 := \min_{k(S)=k}\sigma^2(S)$ is unknown. For a subset $S \subseteq \{1, \ldots, p\}$ of covariates, some might replace $\sigma^2(S)$ with $\hat{\sigma}^2(S)$. However, the value of $\log\hat{\sigma}^2(S)$ is smaller on average than $\log\sigma^2(S)$. In fact, we have the following proposition.

Proposition 19 Let $k(S)$ be the cardinality of S. Then, we have that[3]

$$E[\log\hat{\sigma}^2(S)] = \log\sigma^2(S) - \frac{k(S) + 2}{N} + O\left(\frac{1}{N^2}\right) .$$

---

[3]By $O(f(N))$, we denote a function such that $g(N)/f(N)$ is bounded.

For the proof, see the Appendix at the end of this chapter.

Since, up to $O(N^{-2})$, we have

$$E\left[\log \hat{\sigma}_k^2 + \frac{k+2}{N}\right] = \log \sigma_k^2 \,,$$

the AIC replaces $\log \sigma_k^2$ in (5.13) with $\log \hat{\sigma}_k^2 + \dfrac{k}{N}$ and chooses the k that minimizes

$$N \log \hat{\sigma}_k^2 + 2k \,. \tag{5.14}$$

Appendix: Proof of Propositions

Proposition 18 *For covariates $x_1, \ldots, x_N$, if the responses are $z_1, \ldots, z_N$, the likelihood $-\displaystyle\sum_{i=1}^{N} \log f(z_i|x_i, \gamma)$ of $\gamma \in \mathbb{R}^{p+1}$ is*

$$\frac{N}{2} \log 2\pi\sigma^2 + \frac{1}{2\sigma^2}\|z - X\beta\|^2 - \frac{1}{\sigma^2}(\gamma - \beta)^T X^T (z - X\beta) + \frac{1}{2\sigma^2}(\gamma - \beta)^T X^T X(\gamma - \beta) \tag{5.15}$$

for an arbitrary $\beta \in \mathbb{R}^{p+1}$.

Proof In fact, for $u \in \mathbb{R}$ and $x \in \mathbb{R}^{p+1}$, we have that

$$\log f(u|x, \gamma) = -\frac{1}{2}\log 2\pi\sigma^2 - \frac{1}{2\sigma^2}(u - x\gamma)^2$$

$$(u - x\gamma)^2 = \{(u - x\beta) - x(\gamma - \beta)\}^2$$

$$= (u - x\beta)^2 - 2(\gamma - \beta)^T x^T (u - x\beta) + (\gamma - \beta)^T x^T x(\gamma - \beta)$$

$$\log f(u|x, \gamma) = -\frac{1}{2}\log 2\pi\sigma^2 - \frac{1}{2\sigma^2}(u - x\beta)^2$$

$$+ \frac{1}{\sigma^2}(\gamma - \beta)^T x^T (u - x\beta) - \frac{1}{2\sigma^2}(\gamma - \beta)^T x^T x(\gamma - \beta)$$

and, if we sum over $(x, u) = (x_1, z_1), \ldots, (x_n, z_n)$, we can write

$$-\sum_{i=1}^{N} \log f(z_i|x_i, \gamma) = \frac{N}{2}\log 2\pi\sigma^2 + \frac{1}{2\sigma^2}\|z - X\beta\|^2$$

$$- \frac{1}{\sigma^2}(\gamma - \beta)^T X^T (z - X\beta) + \frac{1}{2\sigma^2}(\gamma - \beta)^T X^T X(\gamma - \beta) \,,$$

where we have used $z = [z_1, \ldots, z_N]^T$ and $\|z - X\beta\|^2 = \sum_{i=1}^{N}(z_i - x_i\beta)^2$, $X^T X =$

$$\sum_{i=1}^{N} x_i^T x_i, \quad X^T(z - X\beta) = \sum_{i=1}^{N} x_i^T(z_i - x_i\beta).$$ $\square$

Proposition 19 *Let $k(S)$ be the cardinality of S. Then, we have*[4]

$$E[\log \hat{\sigma}^2(S)] = \log \sigma^2(S) - \frac{k(S) + 2}{N} + O\left(\frac{1}{N^2}\right).$$

Proof Let $m \geq 1$, $U \sim \chi_m^2$, $V_1, \ldots, V_m \sim N(0, 1)$. For $i = 1, \ldots, m$, we have that

$$Ee^{tV_i^2} = \int_{-\infty}^{\infty} e^{tv_i^2} \frac{1}{\sqrt{2\pi}} e^{-v_i^2/2} dv_i = \int_{-\infty}^{\infty} \frac{1}{\sqrt{2\pi}} \exp\left\{-\frac{(1-2t)v_i^2}{2}\right\} dv_i = (1-2t)^{-1/2}$$

$$Ee^{tU} = \int_{-\infty}^{\infty} e^{t(v_1^2+\cdots+v_m^2)} \frac{1}{\sqrt{2\pi}} e^{-(v_1^2+\cdots+v_m^2)/2} dv_1 \cdots dv_m = (1-2t)^{-m/2},$$

which means that for $n = 1, 2, \ldots$,

$$EU^n = \frac{d^n Ee^{tU}}{dt^n}\Big|_{t=0} = m(m+2)\cdots(m+2n-2),$$ (5.16)

where $Ee^{tU} = 1 + tE[U] + \frac{t^2}{2}E[U^2] + \cdots$ has been used. Moreover, from the Taylor expansion, we have that

$$E\left[\log\frac{U}{m}\right] = E\left(\frac{U}{m} - 1\right) - \frac{1}{2}E\left(\frac{U}{m} - 1\right)^2 + \cdots.$$ (5.17)

If we let (5.16) for $n = 1, 2$, where $EU = m$ and $EU^2 = m(m + 2)$, the first and second terms of (5.17) are zero and

$$-\frac{1}{2m^2}(EU^2 - 2mEU + m^2) = -\frac{1}{2m^2}\{m(m+2) - 2m^2 + m^2\} = -\frac{1}{m},$$

respectively.

---

[4]By $O(f(N))$, we denote a function such that $g(N)/f(N)$ is bounded.

Next, we show that each term in (5.17) for $n \geq 3$ is at most $O(1/m^2)$. From the binomial theorem and (5.16), we have that

$$E(U - m)^n = \sum_{j=0}^{n} \binom{n}{j} EU^j (-m)^{n-j}$$

$$= \sum_{j=0}^{n} (-1)^{n-j} \binom{n}{j} m^{n-j} m(m+2) \cdots (m+2j-2) . \qquad (5.18)$$

If we regard

$$m^{n-j} m(m+2) \cdots (m+2j-2)$$

as a polynomial w.r.t. m, the coefficients of the highest and $(n-1)$-th terms are one and $2\{1 + 2 + \cdots + (j-1)\} = j(j-1)$, respectively. Hence, the coefficients of the n-th and $(n-1)$-th terms in (5.18) are

$$\sum_{j=0}^{n} \binom{n}{j} (-1)^{n-j} = \sum_{j=0}^{n} \binom{n}{j} (-1)^j 1^{n-j} = (-1 + 1)^n = 0$$

and

$$\sum_{j=0}^{n} \binom{n}{j} (-1)^j j(j-1) = \sum_{j=2}^{n} \frac{n!}{(n-j)!(j-2)!} (-1)^{j-2}$$

$$= n(n-1) \sum_{i=0}^{n-2} \binom{n-2}{i} (-1)^i = 0 ,$$

respectively. Thus, we have shown that for $n \geq 3$,

$$E\left(\frac{U}{m} - 1\right)^n = O\left(\frac{1}{m^2}\right) .$$

Finally, from $\dfrac{RSS(S)}{\sigma^2(S)} = \dfrac{N\hat{\sigma}^2(S)}{\sigma^2(S)} \sim \chi^2_{N-k(S)-1}$ and (5.17), if we apply $m = N - k(S) - 1$, then we have that

$$\log \frac{N}{N - k(S) - 1} = \frac{k(S) + 1}{N - k(S) - 1} + O\left(\left(\frac{1}{N - k(S) - 1}\right)^2\right)$$

$$E\left[\log\left(\frac{\hat{\sigma}^2(S)}{N - k(S) - 1} \Big/ \frac{\sigma^2}{N}\right)\right] = -\frac{1}{N - k(S) - 1} + O\left(\frac{1}{N^2}\right) = -\frac{1}{N} + O\left(\frac{1}{N^2}\right)$$

and

$$E\left[\log\frac{\hat{\sigma}^2(S)}{\sigma^2}\right] = -\frac{1}{N} - \frac{k(S)+1}{N} + O\left(\frac{1}{N^2}\right) = -\frac{k(S)+2}{N} + O\left(\frac{1}{N^2}\right).$$

□

Exercises 40–48

In the following, we define

$$X = \begin{bmatrix} x_1 \\ \vdots \\ x_N \end{bmatrix} \in \mathbb{R}^{N\times(p+1)}, \ y = \begin{bmatrix} y_1 \\ \vdots \\ y_N \end{bmatrix} \in \mathbb{R}^N, \ z = \begin{bmatrix} z_1 \\ \vdots \\ z_N \end{bmatrix} \in \mathbb{R}^N, \ \beta = \begin{bmatrix} \beta_0 \\ \beta_1 \\ \vdots \\ \beta_p \end{bmatrix} \in \mathbb{R}^{p+1},$$

where $x_1, \ldots, x_N$ are row vectors. We assume that $X^T X$ has an inverse matrix and denote by $E[\cdot]$ the expectation w.r.t.

$$f(y_i|x_i, \beta) := \frac{1}{\sqrt{2\pi\sigma^2}} \exp\left\{-\frac{\|y_i - x_i\beta\|^2}{2\sigma^2}\right\}.$$

40. For $X \in \mathbb{R}^{N\times(p+1)}$ and $y \in \mathbb{R}^N$, show each of the following:

(a) If the variance $\sigma^2 > 0$ is known, the $\beta \in \mathbb{R}^{p+1}$ that maximizes $l :=$ $\sum_{i=1}^{N} \log f(y_i|x_i, \beta)$ coincides with the least squares solution. Hint:

$$l = -\frac{N}{2}\log(2\pi\sigma^2) - \frac{1}{2\sigma^2}\|y - X\beta\|^2.$$

(b) If both $\beta \in \mathbb{R}^{p+1}$ and $\sigma^2 > 0$ are unknown, the maximum likelihood estimate of σ^2 is given by

$$\hat{\sigma}^2 = \frac{1}{N}\|y - X\hat{\beta}\|^2.$$

Hint: If we partially differentiate l with respect to σ^2, we have

$$\frac{\partial l}{\partial \sigma^2} = -\frac{N}{2\sigma^2} + \frac{\|y - X\beta\|^2}{2(\sigma^2)^2} = 0.$$

(c) For probabilistic density functions f and g over $\mathbb{R}$, the Kullback–Leibler divergence is nonnegative, i.e.,

$$D(f \| g) := \int_{-\infty}^{\infty} f(x) \log \frac{f(x)}{g(x)} dx \geq 0.$$

41. Let $f^N(y|x, \beta) := \prod_{i=1}^{N} f(y_i|x_i, \beta)$. By showing (a) through (d), prove

$$J = \frac{1}{N} E(\nabla l)^2 = -\frac{1}{N} E \nabla^2 l.$$

(a) $\nabla l = \dfrac{\nabla f^N(y|x, \beta)}{f^N(y|x, \beta)}$;

(b) $\displaystyle\int \nabla f^N(y|x, \beta) dy = 0$;

(c) $E \nabla l = 0$;

(d) $\nabla E[\nabla l] = E[\nabla^2 l] + E[(\nabla l)^2]$.

42. Let $\tilde{\beta} \in \mathbb{R}^{p+1}$ be an arbitrary unbiased estimate β. By showing (a) through (c), prove Cramer–Rao's inequality

$$V(\tilde{\beta}) \geq (NJ)^{-1}.$$

(a) $E[(\tilde{\beta} - \beta)(\nabla l)^T] = I$.

(b) The covariance matrix of the vector combining $\tilde{\beta} - \beta$ and ∇l of size $2(p+1)$

$$\begin{bmatrix} V(\tilde{\beta}) & I \\ I & NJ \end{bmatrix}.$$

(c) Both sides of

$$\begin{bmatrix} V(\tilde{\beta}) - (NJ)^{-1} & 0 \\ 0 & NJ \end{bmatrix} = \begin{bmatrix} I & -(NJ)^{-1} \\ 0 & I \end{bmatrix} \begin{bmatrix} V(\tilde{\beta}) & I \\ I & NJ \end{bmatrix} \begin{bmatrix} I & 0 \\ -(NJ)^{-1} & I \end{bmatrix}$$

are nonnegative definite.

43. By showing (a) through (c), prove $E \| X(\tilde{\beta} - \beta) \|^2 \geq \sigma^2(p + 1)$.

(a) $E[(\tilde{\beta} - \beta)^T \nabla l] = p + 1$;

(b) $E \| X(X^T X)^{-1} \nabla l \|^2 = (p + 1)/\sigma^2$;

(c) $\{E(\tilde{\beta} - \beta)^T \nabla l\}^2 \leq E \| X(X^T X)^{-1} \nabla l \|^2 E \| X(\tilde{\beta} - \beta) \|^2$. Hint: For random variables $U, V \in \mathbb{R}^m$ ($m \geq 1$), prove $\{E[U^T V]\}^2 \leq E[\|U\|^2] E[\|V\|^2]$ (Schwarz's inequality).

44. Prove the following statements:

   (a) For covariates $x_1, \ldots, x_N$, if we obtain the responses $z_1, \ldots, z_N$, then the likelihood $-\sum_{i=1}^{N} \log f(z_i|x_i, \gamma)$ of the parameter $\gamma \in \mathbb{R}^{p+1}$ is

$$\frac{N}{2} \log 2\pi\sigma^2 + \frac{1}{2\sigma^2}\|z-X\beta\|^2 - \frac{1}{\sigma^2}(\gamma-\beta)^T X^T (z-X\beta) + \frac{1}{2\sigma^2}(\gamma-\beta)^T X^T X(\gamma-\beta)$$

   for an arbitrary $\beta \in \mathbb{R}^{p+1}$.

   (b) If we take the expectation of (a) w.r.t. $z_1, \ldots, z_N$, it is

$$\frac{N}{2} \log(2\pi\sigma^2 e) + \frac{1}{2\sigma^2}\|X(\gamma - \beta)\|^2 .$$

   (c) If we estimate β and choose an estimate γ of β, the minimum value of (b) on average is

$$\frac{N}{2} \log(2\pi\sigma^2 e) + \frac{1}{2}(p + 1),$$

   and the minimum value is realized by the least squares method.

   (d) Instead of choosing all the p covariates, we choose $0 \le k \le p$ covariates from p. Minimizing

$$\frac{N}{2} \log(2\pi\sigma_k^2 e) + \frac{1}{2}(k + 1)$$

   w.r.t. k is equivalent to minimizing $N \log \sigma_k^2 + k$ w.r.t. k, where σ_k^2 is the minimum variance when we choose k covariates.

45. By showing (a) through (f), prove

$$E \log \frac{\hat{\sigma}^2(S)}{\sigma^2} = -\frac{1}{N} - \frac{k(S) + 1}{N} + O\left(\frac{1}{N^2}\right) = -\frac{k(S) + 2}{N} + O\left(\frac{1}{N^2}\right) .$$

   Use the fact that the moment of $U \sim \chi_m^2$ is

$$EU^n = m(m + 2) \cdots (m + 2n - 2)$$

   without proving it.

   (a) $E \log \dfrac{U}{m} = E\left(\dfrac{U}{m} - 1\right) - \dfrac{1}{2}E\left(\dfrac{U}{m} - 1\right)^2 + \cdots$

   (b) $E\left(\dfrac{U}{m} - 1\right) = 0$ and $E\left(\dfrac{U}{m} - 1\right)^2 = \dfrac{2}{m}$.

(c) $\displaystyle\sum_{j=0}^{n}(-1)^{n-j}\binom{n}{j}=0.$

(d) If we regard $E(U-m)^n = \displaystyle\sum_{j=0}^{n}(-1)^{n-j}\binom{n}{j}m^{n-j}m(m+2)\cdots(m+2j-2)$
as a polynomial of degree m, the sum of the terms of degree n is zero. Hint: Use (c).

(e) The sum of the terms of degree $n-1$ is zero. Hint: Derive that the coefficient of degree $n-1$ is $2\{1+2+\cdots+(j-1)\}=j(j-1)$ for each j and that

$$\sum_{j=0}^{n}\binom{n}{j}(-1)^j j(j-1)=0.$$

(f) $E\log\left(\dfrac{\hat\sigma^2(S)}{N-k(S)-1}\Big/\dfrac{\sigma^2}{N}\right)=-\dfrac{1}{N}+O\left(\dfrac{1}{N^2}\right).$

46. The following procedure produces the AIC value. Fill in the blanks and execute the procedure.

```
from sklearn.linear_model import LinearRegression
import itertools  #  # enumerate combinations
```

```
res=LinearRegression()
```

```
def RSS_min(X,y,T):
    S_min=np.inf
    m=len(T)
    for j in range(m):
        q=T[j]
        res.fit(X[:,q],y)
        y_hat=res.predict(X[:,q])
        S=np.linalg.norm(y_hat-y)**2
        if S<S_min:
            S_min=S
            set_q=q
    return(S_min,set_q)
```

```
from sklearn.datasets import load_boston
```

```
boston=load_boston()
X=boston.data[:, [0,2,4,5,6,7,9,10,11,12]]
y=boston.target
```

```
n,p=X.shape
AIC_min=np.inf
for k in range(1,p+1,1):
    T=list(itertools.combinations(range(p),k))
    #   # each column has combinations (k from p)
    S_min,set_q=RSS_min(X,y,T)
    AIC=# blank(1) #
    if AIC<AIC_min:
        AIC_min=# blank(2) #
        set_min=# blank(3) #
print(AIC_min,set_min)
```

47. Instead of AIC, we consider a criterion that minimizes the following quantity
    (Bayesian Information Criterion (BIC)):

$$N \log \hat{\sigma}^2 + k \log N.$$

Replace the associated lines of the AIC procedure above, and name the function
BIC. For the same data, execute BIC. Moreover, construct a procedure to
choose the covariate set that maximizes

$$AR^2 := 1 - \frac{RSS/(N-k-1)}{TSS/(N-1)}$$

(adjusted coefficient of determination), and name the function AR2. For the
same data, execute AR2.

48. We wish to visualize the k that minimizes AIC and BIC. Fill in the blanks and
    execute the procedure.

```
def IC(X,y,k):
    n,p=X.shape
    T=list(itertools.combinations(range(p),k))
    S,set_q=RSS_min(X,y,T)
    AIC=# blank(1) #
    BIC=# blank(2) #
    return {'AIC':AIC,'BIC':BIC}
```

```
AIC_seq=[]; BIC_seq=[]
for k in range(1,p+1,1):
    AIC_seq.append(# blank(3) #)
    BIC_seq.append(# blank(4) #)
x_seq=np.arange(1,p+1,1)
plt.plot(x_seq,AIC_seq,c="red",label="AIC")
plt.plot(x_seq,BIC_seq,c="blue",label="BIC")
plt.xlabel("the_number_of_variables")
plt.ylabel("values_of_AIC/BIC")
plt.title("changes_of_the_number_of_variables_and_AIC_and_BIC")
plt.legend()
```

Chapter 6
Regularization

Abstract In statistics, we assume that the number of samples N is larger than the number of variables p. Otherwise, linear regression will not produce any least squares solution, or it will find the optimal variable set by comparing the information criterion values of the 2^p subsets of the cardinality p. Therefore, it is difficult to estimate the parameters. In such a sparse situation, regularization is often used. In the case of linear regression, we add a penalty term to the squared error to prevent the coefficient value from increasing. When the regularization term is a constant λ times the L1 and L2 norms of the coefficient, the method is called lasso and ridge, respectively. In the case of lasso, as the constant λ increases, some coefficients become 0; finally, all coefficients become 0 when λ is infinity. In that sense, lasso plays a role of model selection. In this chapter, we consider the principle of lasso and compare it with ridge. Finally, we learn how to choose the constant λ.

6.1 Ridge

In linear regression, assuming that the matrix $X^T X$ with $X \in \mathbb{R}^{N \times (p+1)}$ is nonsingular, we derive that $\hat{\beta} = (X^T X)^{-1} X^T y$ minimizes the per-sample squared error $\|y - X\beta\|^2$ for $y \in \mathbb{R}^N$. In the rest of this chapter, without loss of generality, we regularize only the slope, assuming that the intercept is zero and that X is in $\mathbb{R}^{N \times p}$ rather than in $\mathbb{R}^{N \times (p+1)}$.

Although it is unlikely in general settings that the matrix $X^T X$ is singular, even if the determinant is too small, the confidence interval becomes large and an inconvenient situation occurs. To prevent such situations, letting $\lambda \geq 0$ be a constant, we often use ridge to minimize the square error plus the squared norm of β multiplied by λ:

$$L := \frac{1}{N} \|y - X\beta\|^2 + \lambda \|\beta\|_2^2. \tag{6.1}$$

If we differentiate L by β, we obtain

$$0 = -\frac{2}{N}X^T(y - X\beta) + 2\lambda\beta .$$

If $X^T X + \lambda I$ is nonsingular, we have

$$\hat{\beta} = (X^T X + N\lambda I)^{-1}X^T y ,$$

where $X^T X + N\lambda I$ is nonsingular as long as $\lambda > 0$. In fact, since $X^T X$ is non-negative definite, all the eigenvalues $\mu_1, \ldots, \mu_p$ are nonnegative (Proposition 10). Hence, from Proposition 5, we obtain the eigenvalues of $X^T X + N\lambda I$ by

$$\det(X^T X + N\lambda I - tI) = 0 \Longrightarrow t = \mu_1 + N\lambda, \ldots, \mu_p + N\lambda > 0 .$$

Moreover, from Proposition 6, all the eigenvalues being positive means that the product $\det(X^T X + N\lambda I)$ is positive and that the matrix $X^T X + N\lambda I$ is nonsingular (Proposition 1), which is true for any p and N. If $N < p$, the rank of $X^T X \in \mathbb{R}^{p \times p}$ is at most N (Proposition 3), and it is not nonsingular (Proposition 1). Therefore, we have

$$\lambda > 0 \Longleftrightarrow X^T X + N\lambda I \text{ is nonsingular.}$$

For ridge, we implement the following procedure:

```
def ridge(x,y,lam=0): #lam stands for lambda
    X=copy.copy(x)
    n,p=X.shape
    X_bar=np.zeros(p)
    s=np.zeros(p)
    for j in range(p):
        X_bar[j]=np.mean(X[:,j])
    for j in range(p):
        s[j]=np.std(X[:,j])
        X[:,j]=(X[:,j]-X_bar[j])/s[j]
    y_bar=np.mean(y)
    y=y-y_bar
    beta=np.linalg.inv(X.T@X+n*lam*np.eye(p))@X.T@y
    for j in range(p):
        beta[j]=beta[j]/s[j]
    beta_0=y_bar-X_bar.T@beta
    return {'beta':beta,'beta_0':beta_0}
```

Example 48 We store the dataset (US Crime Data: https://web.stanford.edu/~hastie/StatLearnSparsity/data.html) as a text file crime.txt and apply ridge to find the relation between the response and covariates..

1	Response	Total overall reported crime rate per 1 million residents
2		NA
3	Covariate	Annual police funding in $/resident
4	Covariates	% of people 25+ years with 4 yrs. of high school
5	Covariates	% of 16- to 19-year-olds not in high school and not high school graduates
6	Covariates	% of 18- to 24-year-olds in college
7	Covariates	% of 18- to 24-year-olds in college

We execute the function `ridge` via the following procedure:

```
df=np.loadtxt("crime.txt",delimiter="\t")
X=df[:,[i for i in range(2,7)]]
p=X.shape[1]
y=df[:,0]
lambda_seq=np.arange(0,50,0.5)
plt.xlim(0,50)
plt.ylim(-7.5,15)
plt.xlabel("lambda")
plt.ylabel("beta")
labels=[
"annual_police_funding_in_$resident","%_of_people_25_years+_with_4_yrs._of_
   high_school",
"%_of_16_to_19_year-olds_not_in_highschool_and_not_highschool_graduates","%_
   of_18_to_24_year-olds_in_college",
"%_of_18_to_24_year-olds_in_"]

for j in range(p):
    coef_seq=[]
    for l in lambda_seq:
        coef_seq.append(ridge(X,y,l)['beta'][j])
    plt.plot(lambda_seq,coef_seq,label="{}".format(labels[j]))
plt.legend(loc="upper_right")
```

We illustrate how the coefficients change as λ increases in Fig. 6.1.

6.2 Subderivative

We consider optimizing functions that cannot be differentiated. For example, when we find the points x at which a variable function f such as $f(x) = x^3 - 2x + 1$ is maximal and minimal, by differentiating f with respect to x, we can solve equation $f'(x) = 0$. However, what if the absolute function is contained as in $f(x) = x^2 + x + 2|x|$? To this end, we extend the notion of differentiation.

To begin, we assume that f is convex. In general, if $f(\alpha x + (1-\alpha)y) \leq \alpha f(x) + (1 - \alpha)f(y)$ for an arbitrary $0 < \alpha < 1$ and $x, y \in \mathbb{R}$, we say that f is *convex*.[1]

Ridge: The Coefficients for each λ

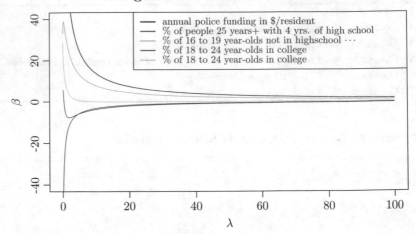

Fig. 6.1 Execution of Example 48. The coefficients β obtained via ridge shrink as λ increases

For example, $f(x) = |x|$ is convex because

$$|\alpha x + (1 - \alpha)y| \leq \alpha|x| + (1 - \alpha)|y| \,.$$

In fact, because both sides are nonnegative, if we subtract the square of the left from that of the right, we have $2\alpha(1 - \alpha)(|xy| - xy) \geq 0$.

If a convex function $f : \mathbb{R} \to \mathbb{R}$ and $x_0 \in \mathbb{R}$ satisfy

$$f(x) \geq f(x_0) + z(x - x_0) \tag{6.2}$$

for $x \in \mathbb{R}$, we say that the set of such $z \in \mathbb{R}$ is the *subderivative* of f at x_0. If a convex f is differentiable at x_0, then z consists of one element[2] $f'(x_0)$, which can be shown as follows.

When a convex function f is differentiable at x_0, we have $f(x) \geq f(x_0) + f'(x_0)(x - x_0)$. In fact, we see that the inequality

$$f(\alpha x + (1 - \alpha)x_0) \leq \alpha f(x) + (1 - \alpha)f(x_0)$$

is equivalent to

$$f(x) - f(x_0) \geq \frac{f(x_0 + \alpha(x - x_0)) - f(x_0)}{\alpha(x - x_0)}(x - x_0) \,.$$

---

[1] In this book, convexity always means convex below and does not mean concave (convex above).
[2] In such a case, we do not express the subderivative as $\{f'(x_0)\}$ but as $f'(x_0)$.

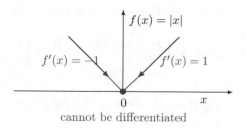

Fig. 6.2 $f(x) = |x|$ cannot be differentiated at the origin. The coefficients from both sides do not match

Then, regardless of $x < x_0$ and $x > x_0$,

$$\frac{f(x_0 + \alpha(x - x_0)) - f(x_0)}{\alpha(x - x_0)}$$

approaches the same $f'(x_0)$ as $\alpha \to 0$. However, we can show that when the convex function f is differentiable at x_0, the z that satisfies (6.2) does not exist, except $f'(x_0)$. In fact, in order for (6.2) to hold for $x > x_0$ and $x < x_0$, we require $\frac{f(x) - f(x_0)}{x - x_0} \geq z$ and $\frac{f(x) - f(x_0)}{x - x_0} \leq z$, respectively, which means that z is no less than the left derivative at x_0 and no more than the right derivative at x_0. Since f is differentiable at $x = x_0$, these values need to coincide.

In this book, we consider only the case of $f(x) = |x|$ and $x_0 = 0$ (Fig. 6.2), and (6.2) becomes $|x| \geq zx$ for an arbitrary $x \in \mathbb{R}$. Then, we can show that the subderivative is the interval $[-1, 1]$:

$$|x| \geq zx , \ x \in \mathbb{R} \Longleftrightarrow |z| \leq 1 .$$

To demonstrate this result, suppose that $|x| \geq zx$ for an arbitrary $x \in \mathbb{R}$. If the claim is true for $x > 0$ and for $x < 0$, we require $z \leq 1$ and $z \geq -1$, respectively. On the other hand, if $-1 \leq z \leq 1$, we have $zx \leq |z| |x| \leq |x|$ for any $x \in \mathbb{R}$.

Example 49 For the cases $x < 0$, $x = 0$, and $x > 0$, we obtain points x such that $f(x) = x^2 - 3x + |x|$ and $f(x) = x^2 + x + 2|x|$ are minimal. For $x \neq 0$, we can differentiate the functions. Note that the subderivative of $f(x) = |x|$ at $x = 0$ is $[-1, 1]$. For the first case

$$f(x) = x^2 - 3x + |x| = \begin{cases} x^2 - 3x + x, \ x \geq 0 \\ x^2 - 3x - x, \ x < 0 \end{cases} = \begin{cases} x^2 - 2x, \ x \geq 0 \\ x^2 - 4x, \ x < 0 \end{cases}$$

$$f'(x) = \begin{cases} 2x - 2, & x > 0 \\ 2x - 3 + [-1, 1] = -3 + [-1, 1] = [-4, -2] \not\ni 0, & x = 0 \\ 2x - 4 < 0, & x < 0 \end{cases}$$

it is minimal at $x = 1$ (Fig. 6.3, left). For the second case

$$f(x) = x^2 + x + 2|x| = \begin{cases} x^2 + x + 2x, \ x \geq 0 \\ x^2 + x - 2x, \ x < 0 \end{cases} = \begin{cases} x^2 + 3x, \ x \geq 0 \\ x^2 - x, \quad x < 0 \end{cases}$$

$$f'(x) = \begin{cases} 2x + 3 > 0, & x > 0 \\ 2x + 1 + 2[-1, 1] = 1 + 2[-1, 1] = [-1, 3] \ni 0, \ x = 0 \\ 2x - 1 < 0, & x < 0 \end{cases}$$

it is minimal at $x = 0$ (Fig. 6.3, right). The graphs are obtained via the following code:

```
x_seq=np.arange(-2,2,0.05)
y=x_seq**2-3*x_seq+np.abs(x_seq)
plt.plot(x_seq,y)
plt.scatter(1,-1,c="red")
plt.title("y=x^2-3x+|x|")
```

```
Text(0.5, 1.0, 'y=x^2-3x+|x|')
```

```
y=x_seq**2+x_seq+2*np.abs(x_seq)
plt.plot(x_seq,y)
plt.scatter(0,0,c="red")
plt.title("y=x^2+x+2|x|")
```

```
Text(0.5, 1.0, 'y=x^2+x+2|x|')
```

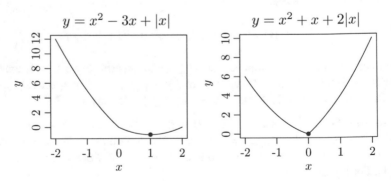

Fig. 6.3 Neither can be differentiated at $x = 0$. The $f(x) = x^2 - 3x + |x|$ (left) and $f(x) = x^2 + x + 2|x|$ (right) are minimal at $x = 1$ and $x = 0$

6.3 Lasso

In ridge, we minimize Eq. (6.1). Lasso also restrains the volume of β, but each coefficient becomes zero when λ exceeds a limit that depends on the coefficient. To examine the mechanism, we replace the second term (L2 norm) $\|\beta\|_2 = \sqrt{\beta_1^2 + \cdots + \beta_p^2}$ in (6.1) with the L1 norm $\|\beta\|_1 = |\beta_1| + \cdots + |\beta_p|$. For $\lambda \geq 0$, we formulate the problem as minimizing

$$L := \frac{1}{2N}\|y - X\beta\|^2 + \lambda\|\beta\|_1 . \tag{6.3}$$

Dividing the first term by two is not essential, and we may double λ if necessary.
For simplicity, we first assume

$$\frac{1}{N}\sum_{i=1}^{N} x_{i,j}x_{i,k} = \begin{cases} 1, & j = k \\ 0, & j \neq k, \end{cases} \tag{6.4}$$

and let $s_j := \dfrac{1}{N}\displaystyle\sum_{i=1}^{N} x_{i,j}y_i$, which will make the derivation easier.

If we differentiate L with respect to β_j, we obtain

$$0 \in -\frac{1}{N}\sum_{i=1}^{N} x_{i,j}\left(y_i - \sum_{k=1}^{p} x_{i,k}\beta_k\right) + \lambda\begin{cases} 1, & \beta_j > 0 \\ -1, & \beta_j < 0 \\ [-1, 1], & \beta_j = 0 \end{cases} \tag{6.5}$$

because the subderivative of $|x|$ at $x = 0$ is $[-1, 1]$. Since we have

$$0 \in \begin{cases} -s_j + \beta_j + \lambda, & \beta_j > 0 \\ -s_j + \beta_j - \lambda, & \beta_j < 0 \\ -s_j + \beta_j + \lambda[-1, 1], & \beta_j = 0, \end{cases}$$

we may write the solution as

$$\beta_j = \begin{cases} s_j - \lambda, & s_j > \lambda \\ s_j + \lambda, & s_j < -\lambda \\ 0, & -\lambda \leq s_j \leq \lambda, \end{cases}$$

where the right-hand side can be expressed as $\beta_j = S_\lambda(s_j)$ with the function

$$S_\lambda(x) = \begin{cases} x - \lambda, & x > \lambda \\ x + \lambda, & x < -\lambda \\ 0, & -\lambda \leq x \leq \lambda. \end{cases}$$

Fig. 6.4 The shape of $\mathcal{S}_\lambda(x)$
for $\lambda = 5$

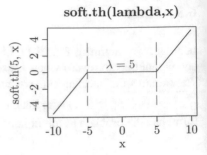

We present the shape of $\mathcal{S}_\lambda(\cdot)$ for $\lambda = 5$ in Fig. 6.4, where we execute the following code:

```
def soft_th(lam,x):
        return np.sign(x)*np.maximum(np.abs(x)-lam,0)
```

```
x_seq=np.arange(-10,10,0.1)
plt.plot(x_seq,soft_th(5,x_seq))
plt.plot([-5,-5],[4,-4],c="black",linestyle="dashed",linewidth=0.8)
plt.plot([5,5],[4,-4],c="black",linestyle="dashed",linewidth=0.8)
plt.title("soft_th(lam,x)")
plt.text(-1.5,1,'lambda=5',fontsize=15)
```

```
Text(-1.5, 1, 'lambda=5')
```

Finally, we remove assumption (6.4). However, relation (6.5) does not hold for this case. To this end, we replace $y_i - \sum_{j=1}^P x_{i,j}\beta_j$ in (6.5) by $r_{i,j} - x_{i,j}\beta_j$ with the residue $r_{i,j} := y_i - \sum_{k \neq j} x_{i,k}\beta_k$ and by $s_j := \dfrac{1}{N}\sum_{i=1}^N r_{i,j}x_{i,j}$ in

$$0 \in -\frac{1}{N}\sum_{i=1}^N x_{i,j}(r_{i,j} - x_{i,j}\beta_j) + \lambda \begin{cases} 1, & \beta_j > 0 \\ -1, & \beta_j < 0 \\ [-1,1], & \beta_j = 0. \end{cases}$$

Then, for fixed β_j, we update β_k for $k \neq j$ and repeat the process for $j = 1, \cdots, p$. We further repeat the cycle until convergence. For example, we can construct the following procedure:

```
def lasso(x,y,lam=0):
    X=copy.copy(x)
    n,p=X.shape
    X_bar=np.zeros(p)
    s=np.zeros(p)
    for j in range(p):
        X_bar[j]=np.mean(X[:,j])
    for j in range(p):
```

```
        s[j]=np.std(X[:,j])
        X[:,j]=(X[:,j]-X_bar[j])/s[j]
    y_bar=np.mean(y)
    y=y-y_bar
    eps=1
    beta=np.zeros(p); beta_old=np.zeros(p)
    while eps>0.001:
        for j in range(p):
            index=list(set(range(p))-{j})
            r=y-X[:,index]@beta[index]
            beta[j]=soft_th(lam,r.T@X[:,j]/n)
        eps=np.max(np.abs(beta-beta_old))
        beta_old=beta
    for j in range(p):
        beta[j]=beta[j]/s[j]
    beta_0=y_bar-X_bar.T@beta
    return {'beta':beta,'beta_0':beta_0}
```

Example 50 We apply the data in Example 48 to lasso.

```
df=np.loadtxt("crime.txt",delimiter="\t")
X=df[:,[i for i in range(2,7,1)]]
p=X.shape[1]
y=df[:,0]
lasso(X,y,20)
```

```
{'beta': array([ 9.65900353, -2.52973842,  3.23224466,  0.        ,  0.        ]),
 'beta_0': 452.208077876934}
```

```
lambda_seq=np.arange(0,200,0.5)
plt.xlim(0,200)
plt.ylim(-10,20)
plt.xlabel("lambda")
plt.ylabel("beta")
labels=["annual_police_funding_in_resident","%_of_people_25_years+_with_4_
    yrs._of_high_school",
"%_of_16_to_19_year-olds_not_in_highschool_and_not_highschool_graduates","%_
    of_18_to_24_year-olds_in_college",
"%_of_18_to_24_year-olds_in_college"]

for j in range(p):
    coef_seq=[]
    for l in lambda_seq:
        coef_seq.append(lasso(X,y,l)['beta'][j])
    plt.plot(lambda_seq,coef_seq,label="{}".format(labels[j]))
plt.legend(loc="upper_right")
plt.title("values_of_each_coefficient_for_each_lambda")
```

```
Text(0.5, 1.0, "values of each coefficient for each lambda")
```

```
lasso(X,y,3.3)
```

```
{'beta': array([10.8009963 , -5.35880785,  4.59591339,  0.13291555,  3.83742115]),
 'beta_0': 497.4278799943754}
```

As shown in Fig. 6.5, the larger the λ is, the smaller the absolute value of the coefficients. We observe that each coefficient becomes zero when λ exceeds a

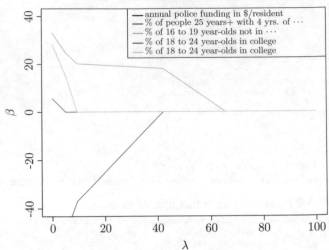

Fig. 6.5 Execution of Example 50. Although the coefficients decrease as λ increases for lasso, each coefficient becomes zero for large λ, and the timing for the coefficients differs

threshold that depends on the coefficient and that the sets of nonzero coefficients depend on the value of λ. The larger the λ is, the smaller the set of nonzero coefficients.

6.4 Comparing Ridge and Lasso

If we compare Figs. 6.1 and 6.5, we find that the absolute values of ridge and lasso decrease as λ increases and that the values approach zero. However, in lasso, each of the coefficients diminishes when λ exceeds a value that depends on the coefficient, and the timing also depends on the coefficients; therefore, we must consider this property for model selection.

Thus far, we have mathematically analyzed ridge and lasso. Additionally, we may wish to intuitively understand the geometrical meaning. Images such as those in Fig. 6.6 are often used to explain the difference between lasso and ridge.

Suppose that $p = 2$ and that $X \in \mathbb{R}^{N \times p}$ consists of two columns $x_{i,1}$ and $x_{i,2}$, $i = 1, \ldots, N$. In the least squares method, we obtain β_1 and β_2 that minimize $S := \sum_{i=1}^{N}(y_i - \beta_1 x_{i,1} - \beta_2 x_{i,2})^2$. Let $\hat{\beta}_1$ and $\hat{\beta}_2$ be the estimates. Since

$$\sum_{i=1}^{N} x_{i,1}(y_i - \hat{y}_i) = \sum_{i=1}^{N} x_{i,2}(y_i - \hat{y}_i) = 0$$

with $\hat{y}_i = \hat{\beta}_1 x_{i1} + \hat{\beta}_2 x_{i2}$ and

$$y_i - \beta_1 x_{i,1} - \beta_2 x_{i,2} = y_i - \hat{y}_i - (\beta_1 - \hat{\beta}_1) x_{i,1} - (\beta_2 - \hat{\beta}_2) x_{i,2}$$

for arbitrary β_1, β_2, the RSS $\sum_{i=1}^{N} (y_i - \beta_1 x_{i,1} - \beta_2 x_{i,2})^2$ can be expressed as

$$(\beta_1 - \hat{\beta}_1)^2 \sum_{i=1}^{N} x_{i,1}^2 + 2(\beta_1 - \hat{\beta}_1)(\beta_2 - \hat{\beta}_2) \sum_{i=1}^{N} x_{i,1} x_{i,2}$$

$$+ (\beta_2 - \hat{\beta}_2)^2 \sum_{i=1}^{N} x_{i,2}^2 + \sum_{i=1}^{N} (y_i - \hat{y}_i)^2 . \tag{6.6}$$

If we let $(\beta_1, \beta_2) := (\hat{\beta}_1, \hat{\beta}_2)$, then we obtain the minimum value $(= RSS)$.

However, when $p = 2$, we may regard solving (6.1) and (6.3) in ridge and lasso as obtaining (β_1, β_2) that minimize (6.6) w.r.t. $\beta_1^2 + \beta_2^2 \leq C$, $|\beta_1| + |\beta_2| \leq C'$ for constants $C, C' > 0$, respectively, where the larger the C, C', the smaller the λ is, where we regard $x_{i1}, x_{i2}, y_i, \hat{y}_i, i = 1, \cdots, N$, and $\hat{\beta}_1, \hat{\beta}_2$ as constants.

The elliptic curve in Fig. 6.6(left) has center $(\hat{\beta}_1, \hat{\beta}_2)$, and each of the contours shares the same value of (6.6). If we expand the contours, then we eventually obtain a rhombus at some (β_1, β_2). Such a pair (β_1, β_2) is the solution of lasso. If the rhombus is smaller (if λ is larger), the elliptic curve is more likely to reach one of the four corners of the rhombus, which means that one of β_1 and β_2 becomes zero. However, as shown in Fig. 6.6(right), if we replace the rhombus with a circle (ridge), it is unlikely that one of β_1 and β_2 becomes zero.

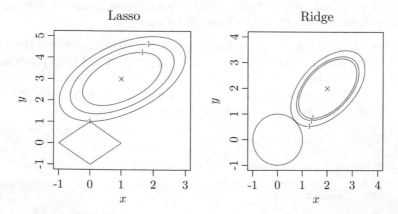

Fig. 6.6 The contours that share the center $(\hat{\beta}_1, \hat{\beta}_2)$ and the square error (6.6), where the rhombus and circle are the constraints of the L1 regularization $|\beta_1| + |\beta_2| \leq C'$ and the L2 regularization $\beta_1^2 + \beta_2^2 \leq C$, respectively

Fig. 6.7 In the green area, the solution satisfies either $\beta_1 = 0$ or $\beta_2 = 0$ when the center is $(\hat{\beta}_1, \hat{\beta}_2)$

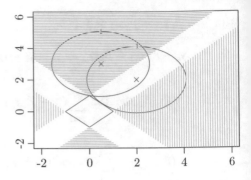

For simplicity, we consider a circle rather than an elliptic curve. In this case, if the solution $(\hat{\beta}_1, \hat{\beta}_2)$ of the least squares method is located somewhere in the green region in Fig. 6.7, either $\beta_1 = 0$ or $\beta_2 = 0$ is the solution. Specifically, if the rhombus is small (λ is large), even if $(\hat{\beta}_1, \hat{\beta}_2)$ remain the same, the area of the green region becomes large.

6.5   Setting the λ Value

When we apply lasso, sklearn.linear_model package `Lasso` is available. Thus far, we have constructed procedures from scratch to understand the principle. We may use the existing package in real applications.

To set the λ value, we usually apply the cross-validation (CV) method in Chap. 3. Suppose that the CV is tenfold. For example, for each λ, we estimate β using nine groups and test the estimate using one group, and we execute this process ten times, changing the groups to evaluate λ. We evaluate all λ values and choose the best. If we input the covariates and response data to the function `LassoCV`, the package evaluates various values of λ and outputs the best one.

Example 51 We apply the data in Examples 48 and 50 to the function `LassoCV` to obtain the best λ. Then, for the best λ, we apply the usual lasso procedure to obtain β. The package outputs the evaluation (the squared error for the test data) and confidence interval for each λ (Fig. 6.8). The numbers on the top of the figure express how many variables are nonzero for the λ value.

```
from sklearn.linear_model import Lasso
from sklearn.linear_model import LassoCV
```

```
Las=Lasso(alpha=20)
Las.fit(X,y)
Las.coef_
```

```
array([11.09067594, -5.2800757 ,  4.65494282,  0.55015932,  2.84324295])
```

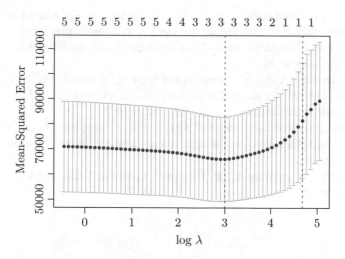

Fig. 6.8 Using the function LassoCV, we obtain the evaluation for each λ (the squared error for the test data), marked as a red point. The vertical segments are the confidence intervals of the true coefficient values. $\log \lambda_{min} = 3$. (The optimum value is approximately $\lambda_{min} = 20$). The numbers on the top of the figure $5, \ldots, 5, 4, \ldots, 4, 3, \ldots, 3, 2, 2, 1, \ldots, 1$ indicate how many variables are nonzero

```
# The grid search for the value specified in alphas
Lcv=LassoCV(alphas=np.arange(0.1,30,0.1),cv=10)
Lcv.fit(X,y)
Lcv.alpha_
Lcv.coef_
```

```
array([11.14516156, -4.87861992,  4.24780979,  0.63662582,  1.52576885])
```

Exercises 49–56

49. Let $N, p \geq 1$. For $X \in \mathbb{R}^{N \times p}$ and $y \in \mathbb{R}^N$, $\lambda \geq 0$, we wish to obtain $\beta \in \mathbb{R}^p$ that minimizes

$$\frac{1}{N} \|y - X\beta\|^2 + \lambda \|\beta\|_2^2 ,$$

where for $\beta = (\beta_1, \ldots, \beta_p)$, we denote $\|\beta\|_2 := \sqrt{\sum_{j=1}^{p} \beta_j^2}$. Suppose $N < p$. Show that such a solution always exists and that it is equivalent to $\lambda > 0$. Hint: In order to show a necessary and sufficient condition, both directions should be proved.

50. (a) Suppose that a function $f : \mathbb{R} \to \mathbb{R}$ is convex and differentiable at $x = x_0$. Show that a z exists for an arbitrary $x \in \mathbb{R}$ such that $f(x) \geq f(x_0) + z(x - x_0)$ (subderivative) and that it coincides with the differential coefficient $f'(x_0)$ at $x = x_0$.

(b) Show that $-1 \leq z \leq 1$ is equivalent to $zx \leq |x|$ for all $x \in \mathbb{R}$.

(c) Find the set of z defined in (a) for function $f(x) = |x|$ and $x_0 \in \mathbb{R}$. Hint: Consider the cases $x_0 > 0$, $x_0 < 0$, and $x_0 = 0$.

(d) Compute the subderivatives of $f(x) = x^2 - 3x + |x|$ and $f(x) = x^2 + x + 2|x|$ for each point, and find the maximal and minimal values for each of the two functions.

51. Write Python program `soft_th(lam,x)` of the function $S_\lambda(x)$, $\lambda > 0$, $x \in \mathbb{R}$ defined by

$$S_\lambda(x) := \begin{cases} x - \lambda, & x > \lambda \\ 0, & |x| \leq \lambda \\ x + \lambda, & x < -\lambda, \end{cases}$$

and execute the following:

```
x_seq=np.arange(-10,10,0.1)
plt.plot(x_seq,soft_th(5,x_seq))
plt.plot([-5,-5],[4,-4],c="black",linestyle="dashed",linewidth=0.8)
plt.plot([5,5],[4,-4],c="black",linestyle="dashed",linewidth=0.8)
plt.title("soft_th(lam,x)")
plt.text(-1.5,1,'lambda=5',fontsize=15)
```

52. We wish to find the $\beta \in \mathbb{R}$ that minimizes

$$L = \frac{1}{2N} \sum_{i=1}^{N} (y_i - x_i\beta)^2 + \lambda|\beta|,$$

given $(x_i, y_i) \in \mathbb{R} \times \mathbb{R}$, $i = 1, \ldots, N$, $\lambda > 0$, where we assume that $x_1, \ldots, x_N$ have been scaled so that $\frac{1}{N} \sum_{i=1}^{N} x_i^2 = 1$. Express the solution by $z := \frac{1}{N} \sum_{i=1}^{N} x_i y_i$ and function $S_\lambda(\cdot)$.

53. For $p > 1$ and $\lambda > 0$, we estimate the coefficients $\beta_0 \in \mathbb{R}$ and $\beta \in \mathbb{R}^p$ as follows: initially, we randomly give the coefficients $\beta \in \mathbb{R}^p$. Then, we update β_j by $S_\lambda \left(\sum_{i=1}^{N} \frac{x_{i,j} r_{i,j}}{N} \right)$, where $r_{i,j} := y_i - \sum_{k \neq j} x_{i,j}\beta_j$. We repeat this process for $j = 1, \ldots, p$ and repeat the cycle until convergence. The function `lasso` below is used to scale the sample-based variance to one for each of the p variables before estimation of (β_0, β). Fill in the blanks and execute the procedure.

```
def lasso(x,y,lam=0): #lam stands for lambda
    X=copy.copy(x)
    n,p=X.shape
```

```
        X_bar=np.zeros(p)
        s=np.zeros(p)
        for j in range(p):
            X_bar[j]=np.mean(X[:,j])
        for j in range(p):
            s[j]=np.std(X[:,j])
            X[:,j]=(X[:,j]-X_bar[j])/s[j]
        y_bar=np.mean(y)
        y=y-y_bar
        eps=1
        beta=np.zeros(p); beta_old=np.zeros(p)
        while eps>0.001:
            for j in range(p):
                index=list(set(range(p))-{j})
                r=# blank(1) #
                beta[j]=# blank(2) #
            eps=np.max(np.abs(beta-beta_old))
            beta_old=beta
        for j in range(p):
            beta[j]=beta[j]/s[j]
        beta_0=# blank(3) #
        return {'beta':beta,'beta_0':beta_0}
```

```
df=np.loadtxt("crime.txt",delimiter="\t")
X=df[:,[i for i in range(2,7,1)]]
p=X.shape[1]
y=df[:,0]
```

```
lambda_seq=np.arange(0,200,0.5)
plt.xlim(0,200)
plt.ylim(-7.5,15)
plt.xlabel("lambda")
plt.ylabel("beta")
labels=["annual_police_funding_in_resident","%_of_people_25_years+_with_
    4_yrs._of_high_school",
"%_of_16_to_19_year-olds_not_in_highschool_and_not_highschool_graduates"
    ,"%_of_18_to_24_year-olds_in_college",
"%_of_18_to_24_year-olds_in_college"]

for j in range(p):
    coef_seq=[]
    for l in lambda_seq:
        coef_seq.append(# blank(4) #)
    plt.plot(lambda_seq,coef_seq,label="{}".format(labels[j]))
plt.legend(loc="upper_right")
plt.title("values_of_each_coefficient_for_each_lambda")
```

54. Transform Problem 53(lasso) into the setting in Problem 49(ridge) and execute
    it. Hint: Replace the line of eps and the while loop in the function lasso
    by

```
beta=np.linalg.inv(X.T@X+n*lam*np.eye(p))@X.T@y
```

and change the function name to ridge. Blank (4) should be ridge rather
than lasso.

55. Look up the meanings of `Lasso` and `LassoCV` and find the optimal λ and β for the data below. Which variables are selected among the five variables?

```
from sklearn.linear_model import Lasso
```

```
Las=Lasso(alpha=20)
Las.fit(X,y)
Las.coef_
```

```
array([132.15580773,  -24.96440514,   19.26809441,   0.
                0.          ])
```

```
#  The grid search for the value specified in alphas
Lcv=LassoCV(alphas=np.arange(0.1,30,0.1),cv=10)
Lcv.fit(X,y)
Lcv.alpha_
Lcv.coef_
```

Hint: The coefficients are displayed via `Lcv.coef_`. If a coefficient is nonzero, we consider it to be selected.

56. Given $x_{i,1}, x_{i,2}, y_i \in \mathbb{R}$, $i = 1, \dots, N$, let $\hat{\beta}_1$ and $\hat{\beta}_2$ be the β_1 and β_2 that minimize $S := \sum_{i=1}^{N}(y_i - \beta_1 x_{i,1} - \beta_2 x_{i,2})^2$ given $\hat{\beta}_1 x_{i,1} + \hat{\beta}_2 x_{i,2}, \hat{y}_i, (i = 1, \dots, N)$. Show the following three equations:

(a)

$$\sum_{i=1}^{N} x_{i,1}(y_i - \hat{y}_i) = \sum_{i=1}^{N} x_{i,2}(y_i - \hat{y}_i) = 0 .$$

For arbitrary β_1, β_2,

$$y_i - \beta_1 x_{i,1} - \beta_2 x_{i,2} = y_i - \hat{y}_i - (\beta_1 - \hat{\beta}_1)x_{i,1} - (\beta_2 - \hat{\beta}_2)x_{i,2} .$$

For arbitrary β_1, β_2, $\sum_{i=1}^{N}(y_i - \beta_1 x_{i,1} - \beta_2 x_{i,2})^2$ can be expressed by

$$(\beta_1 - \hat{\beta}_1)^2 \sum_{i=1}^{N} x_{i,1}^2 \ + 2(\beta_1 - \hat{\beta}_1)(\beta_2 - \hat{\beta}_2) \sum_{i=1}^{N} x_{i,1}x_{i,2} + (\beta_2 - \hat{\beta}_2)^2 \sum_{i=1}^{N} x_{i,2}^2$$

$$+ \sum_{i=1}^{N}(y_i - \hat{y}_i)^2.$$

(b) We consider the case $\sum_{i=1}^{N} x_{i,1}^2 = \sum_{i=1}^{N} x_{i,2}^2 = 1$ and $\sum_{i=1}^{N} x_{i,1} x_{i,2} = 0$. In
the standard least squares method, we choose the coefficients as $\beta_1 = \hat{\beta}_1$
and $\beta_2 = \hat{\beta}_2$. However, under the constraint that $|\beta_1| + |\beta_2|$ is less than a
constant, we choose (β_1, β_2) at which the circle with center $(\hat{\beta}_1, \hat{\beta}_2)$ and the
smallest radius comes into contact with the rhombus. Suppose that we grow
the radius of the circle with center $(\hat{\beta}_1, \hat{\beta}_2)$ until it comes into contact with
the rhombus that connects $(1, 0)$, $(0, 1)$, $(-1, 0)$, $(0, -1)$. Show the region
of the centers such that one of the coordinates $(\hat{\beta}_1$ and $\hat{\beta}_2)$ is zero.

(c) What if the rhombus in (b) is replaced by a unit circle?

Chapter 7
Nonlinear Regression

Abstract For regression, until now we have focused on only linear regression, but in this chapter, we will consider the nonlinear case where the relationship between the covariates and response is not linear. In the case of linear regression in Chap. 2, if there are p variables, we calculate $p + 1$ coefficients of the basis that consists of $p + 1$ functions $1, x_1, \cdots, x_p$. This chapter addresses regression when the basis is general. For example, if the response is expressed as a polynomial of the covariate x, the basis consists of $1, x, \cdots, x^p$. We also consider spline regression and find a basis. In that case, the coefficients can be found in the same manner as for linear regression. Moreover, we consider local regression for which the response cannot be expressed by a finite number of basis functions. Finally, we consider a unified framework (generalized additive model) and backfitting.

7.1 Polynomial Regression

We consider fitting the relation between the covariates and response to a polynomial from observed data $(x_1, y_1), \ldots, (x_N, y_N) \in \mathbb{R} \times \mathbb{R}$. By a polynomial, we mean the function $f : \mathbb{R} \to \mathbb{R}$ that is determined by specifying the coefficients $\beta_0, \beta_1, \ldots, \beta_p$ in $\beta_0 + \beta_1 x + \cdots + \beta_p x^p$ for $p \geq 1$, such as $f(x) = 1 + 2x - 4x^3$. As we do in the least squares method, we assume that the coefficients $\beta_0, \ldots, \beta_p$ minimize

$$\sum_{i=1}^{N} (y_i - \beta_0 - \beta_1 x_i - \cdots - \beta_p x_i^p)^2 .$$

By overlapping $x_{i,j}$ and x_i^j, if the matrix $X^T X$ is nonsingular with

$$X = \begin{bmatrix} 1 & x_1 & \cdots & x_1^p \\ \vdots & \vdots & \ddots & \vdots \\ 1 & x_N & \cdots & x_N^p \end{bmatrix},$$

J. Suzuki, *Statistical Learning with Math and Python*,
https://doi.org/10.1007/978-981-15-7877-9_7

Fig. 7.1 We generated the
data by adding standard
Gaussian random values to a
sine curve and fit the data to
polynomials of orders
$p = 3, 5, 7$

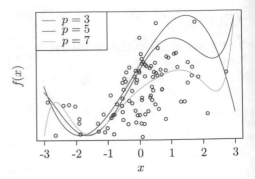

we can check that $\hat{\beta} = (X^T X)^{-1} X^T y$ is the solution. As in linear regression $\hat{f}(x) = \hat{\beta}_0 + \hat{\beta}_1 x_1 + \cdots + \hat{\beta}_p x_p$, from the obtained $\hat{\beta}_0, \ldots, \hat{\beta}_p$, we construct an estimated function

$$\hat{f}(x) = \hat{\beta}_0 + \hat{\beta}_1 x + \cdots + \hat{\beta}_p x^p .$$

Example 52 We generate $N = 100$ observed data by adding standard Gaussian random values to the sine function and fit them to polynomials of orders $p = 3, 5, 7$. We show the results in Fig. 7.1. The generation of polynomials is achieved via the following code:

```
def g(beta,u):
    S=0
    for j in range(p+1):   # length of beta = p+1
        S=S+beta[j]*u**j
    return S
```

```
n=100; x=randn(n); y=np.sin(x)+randn(n)
m=3
p_set=[3,5,7]
col_set=["red","blue","green"]
randn(3)*randn(3)**np.array([1,2,3])
```

```
    array([ 0.07981705, -0.06213429, -0.01101873])
```

```
plt.scatter(x,y,s=20,c="black")
plt.ylim(-2.5,2.5)
x_seq=np.arange(-3,3,0.1)
for i in range(m):
    p=p_set[i]
    X=np.ones([n,1])
    for j in range(1,p+1):
        xx=np.array(x**j).reshape((n,1))
        X=np.hstack((X,xx))
    beta=np.linalg.inv(X.T@X)@X.T@y
    def f(u):
        return g(beta,u)
```

```
    plt.plot(x_seq,f(x_seq),c=col_set[i],label="p={}".format(p))
plt.legend(loc="lower_right")
```

We can show that if no less than $p+1$ are different among $x_1, \ldots, x_N$, the matrix $X^T X$ is nonsingular. To examine this claim, since the ranks of $X^T X$ and X are equal (see Sect. 2.2), it is sufficient to show that the determinant is not zero for the matrix such that the $p+1$ columns are contained in $X \in \mathbb{R}^{N \times (p+1)}$, which is true from the fact that, in Example 7, the determinant of the $n \times n$ Vandermonde's matrix is not zero if $a_1, \ldots, a_n$ are different.

Polynomial regression can be applied to more general settings. For $f_0 = 1$ and $f_1, \ldots, f_p : \mathbb{R} \to \mathbb{R}$, we can compute $\hat{\beta} = (X^T X)^{-1} X^T y$ as long as each of the columns in

$$
X = \begin{bmatrix} 1 & f_1(x_1) & \cdots & f_p(x_1) \\ \vdots & \vdots & \ddots & \vdots \\ 1 & f_1(x_N) & \cdots & f_p(x_N) \end{bmatrix}
$$

is linearly independent. From the obtained $\hat{\beta}_0, \ldots, \hat{\beta}_p$, we can construct

$$
\hat{f}(x) = \hat{\beta}_0 f_0(x) + \hat{\beta}_1 f_1(x) + \cdots + \hat{\beta}_p f_p(x) ,
$$

where we often assume $f_0(x) = 1$.

Example 53 We then generate $x \sim N(0, \pi^2)$ and

$$
y = \begin{cases} -1 + \epsilon, & 2m - 1 \le |x| < 2m \\ 1 + \epsilon, & 2m - 2 \le |x| < 2m - 1 \end{cases}, \quad m = 1, 2, \ldots \tag{7.1}
$$

(Fig. 7.2), where $\epsilon \sim N(0, 0.2^2)$. We observe that the even functions $f_1(x) = 1$, $f_2(x) = \cos x$, $f_3(x) = \cos 2x$, and $f_4(x) = \cos 3x$ are better to fit than the odd functions $f_1(x) = 1$, $f_2(x) = \sin x$, $f_3(x) = \sin 2x$, and $f_4(x) = \sin 3x$ (Fig. 7.3) because we generated the observed data according to an even function with added noise.

Fig. 7.2 The graph of the function obtained by removing noise in (7.1). It is an even and cyclic function

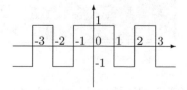

Fig. 7.3 We generated data such that whether y is close to either -1 or 1 is based on whether x is even or odd when truncating it. Note that (7.1) is a cyclic and even function when removing the noise ϵ (Fig. 7.2). We observe that $\cos nx$, $n = 1, 2, \ldots$, are better to fit than $\sin nx$, $n = 1, 2, \ldots$

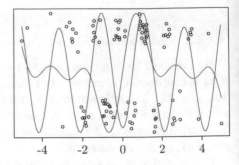

The procedure is implemented via the following code:

```
# Generating data close to an even function
n=100
x=randn(n)*np.pi
y=np.round(x)%2*2-1+randn(n)*0.2

# Write axes, etc.
plt.scatter(x,y,s=20,c="black")
plt.tick_params(labelleft=False)
x_seq=np.arange(-8,8,0.2)
```

```
def f(x,g):
    return beta[0]+beta[1]*g(x)+beta[2]*g(2*x)+beta[3]*g(3*x)
```

```
# select 1, cosx ,cos2x and cos3x as basis
X=np.ones([n,1])
for j in range(1,4):
    xx=np.array(np.cos(j*x)).reshape((n,1))
    X=np.hstack((X,xx))
beta=np.linalg.inv(X.T@X)@X.T@y
plt.plot(x_seq,f(x_seq,np.cos),c="red")
```

```
#  select 1, sinx ,sin2x and sin3x as basis
X=np.ones([n,1])
for j in range(1,4):
    xx=np.array(np.sin(j*x)).reshape((n,1))
    X=np.hstack((X,xx))
beta=np.linalg.inv(X.T@X)@X.T@y
plt.plot(x_seq,f(x_seq,np.sin),c="blue")
```

7.2 Spline Regression

In this section, we restrict the polynomials to those with an order at most three, such as $x^3 + x^2 - 7, -8x^3 - 2x + 1$.

We first note that if polynomials f and g of order $p = 3$ coincide with each other up to the second derivative at the point $x_* \in \mathbb{R}$: $f^{(j)}(x_*) = g^{(j)}(x_*)$, $j = 0, 1, 2$, in

$$
\begin{cases}
f(x) = \displaystyle\sum_{j=0}^{3} \beta_j (x - x_*)^j \\
g(x) = \displaystyle\sum_{j=0}^{3} \gamma_j (x - x_*)^j;
\end{cases}
$$

then, we have $\beta_j = \gamma_j$, $j = 0, 1, 2$. In fact, we see that $f(x_*) = g(x_*)$, $f'(x_*) = g'(x_*)$, and $f''(x_*) = g''(x_*)$ imply $2\beta_2 = 2\gamma_2$, $\beta_1 = \gamma_1$, and $\beta_0 = \gamma_0$, respectively. Hence, we have

$$
f(x) - g(x) = (\beta_3 - \gamma_3)(x - x_*)^3 .
$$

In the following, for $K \geq 1$, we divide the line $\mathbb{R}$ at the knots $-\infty = \alpha_0 < \alpha_1 < \cdots < \alpha_K < \alpha_{K+1} = \infty$ and express the function $f(x)$ as a polynomial $f_i(x)$ for each $\alpha_i \leq x \leq \alpha_{i+1}$, where we assume that those $K + 1$ functions are continuous up to the second derivative at the K knots:

$$
f_{i-1}^{(j)}(\alpha_i) = f_i^{(j)}(\alpha_i), \quad j = 0, 1, 2, \quad i = 1, \ldots, K \tag{7.2}
$$

(*spline function*). Note that there exists a constant γ_i such that $f_i(x) = f_{i-1}(x) + \gamma_i (x - \alpha_i)^3$ for each $i = 1, 2, \ldots, K + 1$.

In (7.2), there are $3K$ linear constraints for $4(K + 1)$ variables w.r.t. $K + 1$ cubic polynomials, each of which contains four coefficients, which means that there remain $K + 4$ degrees of freedom. We first arbitrarily determine the values of $\beta_0, \beta_1, \beta_2$, and β_3 in $f_0(x) = \beta_0 + \beta_1 x + \beta_2 x^2 + \beta_3 x^3$ for $\alpha_0 \leq x \leq \alpha_1$. Next, noting that for each $i = 1, 2, \ldots, K$, the difference between f_i and f_{i-1} is $(x - \alpha_i)^3$ multiplied by a constant β_{i+3}, all the polynomials are determined by specifying β_{i+3}, $i = 1, 2, \ldots, K$. We express the function f as follows:

$$
f(x) =
\begin{cases}
\beta_0 + \beta_1 x + \beta_2 x^2 + \beta_3 x^3, & \alpha_0 \leq x \leq \alpha_1 \\
\beta_0 + \beta_1 x + \beta_2 x^2 + \beta_3 x^3 + \beta_4 (x - \alpha_1)^3, & \alpha_1 \leq x \leq \alpha_2 \\
\beta_0 + \beta_1 x + \beta_2 x^2 + \beta_3 x^3 + \beta_4 (x - \alpha_1)^3 + \beta_5 (x - \alpha_2)^3, & \alpha_2 \leq x \leq \alpha_3 \\
\quad \vdots & \quad \vdots \\
\beta_0 + \beta_1 x + \beta_2 x^2 + \beta_3 x^3 + \beta_4 (x - \alpha_1)^3 \\
\quad + \beta_5 (x - \alpha_2)^3 + \cdots + \beta_{K+3} (x - \alpha_K)^3, & \alpha_K \leq x \leq \alpha_{K+1}
\end{cases}
$$

$$
= \beta_0 + \beta_1 x + \beta_2 x^2 + \beta_3 x^3 + \sum_{i=1}^{K} \beta_{i+3} (x - \alpha_i)_+^3 ,
$$

where $(x - \alpha_i)_+$ is the function that takes $x - \alpha_i$ and zero for $x > \alpha_i$ and for $x \le \alpha_i$, respectively. The method for choosing the coefficients $\beta_0, \ldots, \beta_{K+3}$ is similar to the method we use for linear regression. Suppose we have observations $(x_1, y_1), \ldots, (x_N, y_N)$, where the sample points $x_1, \ldots, x_N$ and the knots $\alpha_1, \ldots, \alpha_K$ should not be confused. For the matrix

$$
X = \begin{bmatrix}
1 & x_1 & x_1^2 & x_1^3 & (x_1 - \alpha_1)_+^3 & (x_1 - \alpha_2)_+^3 & \cdots & (x_1 - \alpha_K)_+^3 \\
1 & x_2 & x_2^2 & x_2^3 & (x_2 - \alpha_1)_+^3 & (x_2 - \alpha_2)_+^3 & \cdots & (x_2 - \alpha_K)_+^3 \\
\vdots & \vdots & \vdots & \vdots & \vdots & \vdots & \vdots & \vdots \\
1 & x_N & x_N^2 & x_N^3 & (x_N - \alpha_1)_+^3 & (x_N - \alpha_2)_+^3 & \cdots & (x_N - \alpha_K)_+^3
\end{bmatrix},
$$

we determine the $\beta = [\beta_0, \ldots, \beta_{K+3}]^T$ that minimizes

$$
\sum_{i=1}^{N} \{y_i - \beta_0 - x_i\beta_1 - x_i^2\beta_2 - x_i^3\beta_3 - (x_i - \alpha_1)_+^3\beta_4 - (x_i - \alpha_2)_+^3\beta_5 - \cdots - (x_i - \alpha_K)_+^3\beta_{K+3}\}^2 .
$$

If the rank is $K + 4$, i.e., the $K + 4$ columns of X are linearly independent, then $X^T X$ is nonsingular, and we obtain the solution $\hat\beta = (X^T X)^{-1} X^T y$ (Fig. 7.4).

Example 54 After generating data, we execute spline regression with $K = 5, 7, 9$ knots. We present the results in Fig. 7.5.

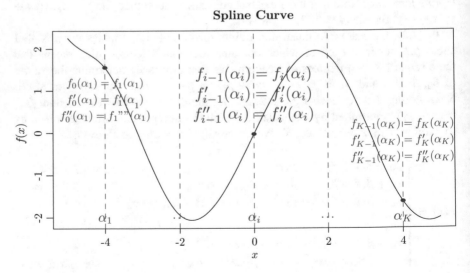

Fig. 7.4 In spline functions, the value and the first and second derivatives should coincide on the left and right of each knot

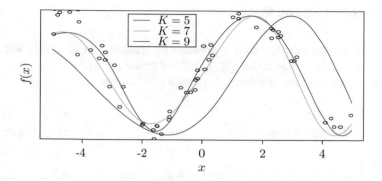

Fig. 7.5 Spline regression with $K = 5, 7, 9$ knots (Example 54)

```
n=100
x=randn(n)*2*np.pi
y=np.sin(x)+0.2*randn(n)
col_set=["red","green","blue"]
K_set=[5,7,9]
plt.scatter(x,y,c="black",s=10)
plt.xlim(-5,5)
for k in range(3):
    K=K_set[k]
    knots=np.linspace(-2*np.pi,2*np.pi,K)
    X=np.zeros((n,K+4))
    for i in range(n):
        X[i,0]=1
        X[i,1]=x[i]
        X[i,2]=x[i]**2
        X[i,3]=x[i]**3
        for j in range(K):
            X[i,j+4]=np.maximum((x[i]-knots[j])**3,0)
    beta=np.linalg.inv(X.T@X)@X.T@y
    def f(x):
        S=beta[0]+beta[1]*x+beta[2]*x**2+beta[3]*x**3
        for j in range(K):
            S=S+beta[j+4]*np.maximum((x-knots[j])**3,0)
        return S
    u_seq=np.arange(-5,5,0.02)
    v_seq=[]
    for u in u_seq:
        v_seq.append(f(u))
    plt.plot(u_seq,v_seq,c=col_set[k],label="K={}".format(K))
plt.legend()
```

7.3   Natural Spline Regression

In this section, we modify spline regression by replacing cubic curves with lines only for both ends $x \leq \alpha_1$ and $\alpha_K \leq x$ (natural spline curve).

Suppose we write the function f for $x \leq \alpha_K$ as follows:

$$f(x) = \begin{cases} \beta_1 + \beta_2 x, & \alpha_0 \leq x \leq \alpha_1 \\ \beta_1 + \beta_2 x + \beta_3 (x - \alpha_1)^3, & \alpha_1 \leq x \leq \alpha_2 \\ \vdots & \vdots \\ \beta_1 + \beta_2 x + \beta_3 (x - \alpha_1)^3 + \cdots + \beta_K (x - \alpha_{K-2})^3, & \alpha_{K-2} \leq x \leq \alpha_{K-1} \\ \beta_1 + \beta_2 x + \beta_3 (x - \alpha_1)^3 + \cdots \\ \quad + \beta_K (x - \alpha_{K-2})^3 + \beta_{K+1} (x - \alpha_{K-1})^3, & \alpha_{K-1} \leq x \leq \alpha_K. \end{cases}$$

Since the second derivative at $x = \alpha_K$ is zero, we have $6 \sum_{j=3}^{K+1} \beta_j (\alpha_K - \alpha_{j-2}) = 0$, and we obtain

$$\beta_{K+1} = -\sum_{j=3}^{K} \frac{\alpha_K - \alpha_{j-2}}{\alpha_K - \alpha_{K-1}} \beta_j . \tag{7.3}$$

Then, if we find the values of $\beta_1, \cdots, \beta_K$, we obtain the values of $f(\alpha_K)$ and $f'(\alpha_K)$ and the line $y = f'(\alpha_K)(x - \alpha_K) + f(\alpha_K)$ for $x \geq \alpha_K$ (Fig. 7.6). Thus, the function f is obtained by specifying $\beta_1, \ldots, \beta_K$.

Proposition 20 *The function $f(x)$ has K cubic polynomials $h_1(x) = 1$, $h_2(x) = x$, $h_{j+2}(x) = d_j(x) - d_{K-1}(x)$, $j = 1, \ldots, K - 2$, as a basis, and if we define*

$$\gamma_1 := \beta_1, \ \gamma_2 := \beta_2, \ \gamma_3 := (\alpha_K - \alpha_1)\beta_3, \ \ldots, \ \gamma_K := (\alpha_K - \alpha_{K-2})\beta_K$$

for each $\beta_1, \ldots, \beta_K$, then we can express f by $f(x) = \sum_{j=1}^{K} \gamma_j h_j(x)$, where we have

$$d_j(x) = \frac{(x - \alpha_j)_+^3 - (x - \alpha_K)_+^3}{\alpha_K - \alpha_j}, \ j = 1, \ldots, K - 1 .$$

For the proof, see the Appendix at the end of this chapter.

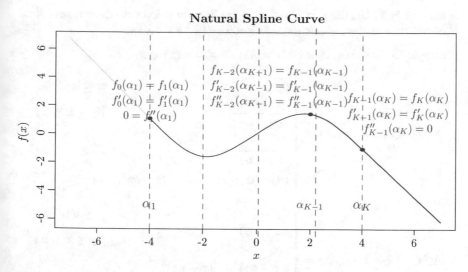

Fig. 7.6 In the natural spline curves, we choose the slope and intercept of the line for $x \leq \alpha_1$ (two degrees of freedom) and the coefficients for $\alpha_i \leq x \leq \alpha_{i+1}$ (one degree of freedom for each $i = 1, 2, \ldots, K - 2$). However, no degrees of freedom are left for $\alpha_{K-1} \leq x \leq \alpha_K$ because $f''(\alpha) = 0$. Moreover, for $\alpha_K \leq x$, the slope and intercept are determined from the values of $f(\alpha_K)$ and $f'(\alpha_K)$, and no degrees of freedom are left as well

We can construct the corresponding Python code as follows:

```python
def d(j,x,knots):
    K=len(knots)
    return (np.maximum((x-knots[j])**3,0)-np.maximum((x-knots[K-1])**3,0))/(
        knots[K-1]-knots[j])
```

```python
def h(j,x,knots):
    K=len(knots)
    if j==0:
        return 1
    elif j==1:
        return x
    else :
        return (d(j-2,x,knots)-d(K-2,x,knots)) # Note that the way of
            counting in array is beginning 0.
```

If we are given observations $(x_1, y_1), \ldots, (x_N, y_N)$, then we wish to determine γ that minimizes $\|y - X\gamma\|^2$ with

$$X = \begin{bmatrix} h_1(x_1) = 1 & h_2(x_1) & \cdots & h_K(x_1) \\ h_1(x_2) = 1 & h_2(x_2) & \cdots & h_K(x_2) \\ \vdots & \vdots & \cdots & \vdots \\ h_1(x_N) = 1 & h_2(x_N) & \cdots & h_K(x_N) \end{bmatrix}. \tag{7.4}$$

If the rank is K, i.e., the K columns in X are linearly independent, the matrix $X^T X$ is nonsingular, and we obtain the solution $\hat{\gamma} = (X^T X)^{-1} X^T y$.

Example 55 If $K = 4$, then we have $h_1(x) = 1$, $h_2(x) = x$,

$$h_3(x) = d_1(x) - d_3(x) = \begin{cases} 0, & x \le \alpha_1 \\ \dfrac{(x - \alpha_1)^3}{\alpha_4 - \alpha_1}, & \alpha_1 \le x \le \alpha_3 \\ \dfrac{(x - \alpha_1)^3}{\alpha_4 - \alpha_1} - \dfrac{(x - \alpha_3)^3}{\alpha_4 - \alpha_3}, & \alpha_3 \le x \le \alpha_4 \\ (\alpha_3 - \alpha_1)(3x - \alpha_1 - \alpha_3 - \alpha_4), & \alpha_4 \le x \end{cases}$$

$$h_4(x) = d_2(x) - d_3(x) = \begin{cases} 0, & x \le \alpha_2 \\ \dfrac{(x - \alpha_2)^3}{\alpha_4 - \alpha_2}, & \alpha_2 \le x \le \alpha_3 \\ \dfrac{(x - \alpha_2)^3}{\alpha_4 - \alpha_2} - \dfrac{(x - \alpha_3)^3}{\alpha_4 - \alpha_3}, & \alpha_3 \le x \le \alpha_4 \\ (\alpha_3 - \alpha_2)(3x - \alpha_2 - \alpha_3 - \alpha_4), & \alpha_4 \le x. \end{cases}$$

Hence, the lines for $x \le \alpha_1$ and $x \ge \alpha_4$ are

$$f(x) = \gamma_1 + \gamma_2 x, \ x \le \alpha_1$$
$$f(x) = \gamma_1 + \gamma_2 x + \gamma_3(\alpha_3 - \alpha_1)(3x - \alpha_1 - \alpha_3 - \alpha_4)$$
$$+ \gamma_4(\alpha_3 - \alpha_2)(3x - \alpha_2 - \alpha_3 - \alpha_4), \ x \ge \alpha_4.$$

Example 56 We compare the ordinary and natural spline curves (Fig. 7.7). By definition, the natural spline becomes a line at both ends, although considerable differences are observed near the points α_1 and α_K. The procedure is implemented according to the following code:

```
n=100
x=randn(n)*2*np.pi
y=np.sin(x)+0.2*randn(n)
K=11
knots=np.linspace(-5,5,K)
X=np.zeros((n,K+4))
for i in range(n):
    X[i,0]=1
    X[i,1]=x[i]
    X[i,2]=x[i]**2
    X[i,3]=x[i]**3
    for j in range(K):
        X[i,j+4]=np.maximum((x[i]-knots[j])**3,0)
beta=np.linalg.inv(X.T@X)@X.T@y
```

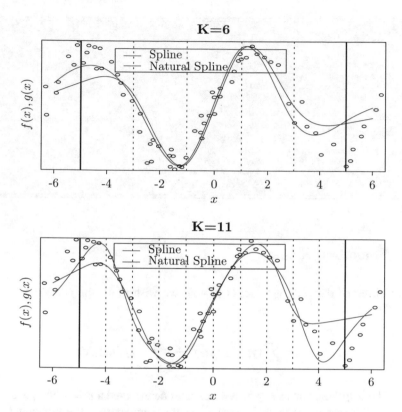

Fig. 7.7 Comparison of the ordinary (blue) and natural (red) splines when $K = 6$ (left) and $K = 11$ (right) in Example 56. While the natural spline becomes a line for each of the both ends, they do not coincide inside the region, in particular, near the borders

```
def f(x):
    S=beta[0]+beta[1]*x+beta[2]*x**2+beta[3]*x**3
    for j in range(K):
        S=S+beta[j+4]*np.maximum((x-knots[j])**3,0)
    return S
```

```
X=np.zeros((n,K))
X[:,0]=1
for j in range(1,K):
    for i in range(n):
        X[i,j]=h(j,x[i],knots)
gamma=np.linalg.inv(X.T@X)@X.T@y
```

```
def g(x):
    S=gamma[0]
    for j in range(1,K):
        S=S+gamma[j]*h(j,x,knots)
    return S
```

```
u_seq=np.arange(-6,6,0.02)
v_seq=[] ; w_seq=[]
for u in u_seq:
    v_seq.append(f(u))
    w_seq.append(g(u))
plt.scatter(x,y,c="black",s=10)
plt.xlim(-6,6)
plt.xlabel("x")
plt.ylabel("f(x),g(x)")
plt.tick_params(labelleft=False)
plt.plot(u_seq,v_seq,c="blue",label="spline_")
plt.plot(u_seq,w_seq,c="red",label="_natural_spline")
plt.vlines(x=[-5,5],ymin=-1.5,ymax=1.5,linewidth=1)
plt.vlines(x=knots,ymin=-1.5,ymax=1.5,linewidth=0.5,linestyle="dashed")
plt.legend()
```

7.4   Smoothing Spline

Given observed data $(x_1, y_1), \ldots, (x_N, y_N)$, we wish to obtain $f : \mathbb{R} \to \mathbb{R}$ that minimizes

$$L(f) := \sum_{i=1}^{N}(y_i - f(x_i))^2 + \lambda \int_{-\infty}^{\infty}\{f''(x)\}^2 dx \qquad (7.5)$$

(*smoothing spline*), where $\lambda \geq 0$ is a constant determined a priori. Suppose $x_1 < \cdots < x_N$. The second term in (7.5) penalizes the complexity of the function f, and $\{f''(x)\}^2$ intuitively expresses how nonsmooth the function is at x. If f is linear, the value is zero, and if λ is small, although the curve meanders, the curve is easier to fit to the observed data. On the other hand, if λ is large, although the curve does not follow the observed data, the curve is smoother.

First, we show that the optimal f is realized by the natural spline with knots $x_1, \ldots, x_N$.

Proposition 21 (Green and Silverman, 1994) *The natural spline f with knots $x_1, \ldots, x_N$ minimizes $L(f)$.*

See the Appendix at the end of this chapter for the proof.

Next, we obtain the coefficients $\gamma_1, \ldots, \gamma_N$ of such a natural spline $f(x) = \sum_{i=1}^{N} \gamma_i h_i(x)$. Let $G = (g_{i,j})$ be the matrix with elements

$$g_{i,j} := \int_{-\infty}^{\infty} h_i''(x)h_j''(x)dx . \qquad (7.6)$$

Then, the second term in $L(g)$ becomes

$$\lambda \int_{-\infty}^{\infty} \{f''(x)\}^2 dx = \lambda \int_{-\infty}^{\infty} \sum_{i=1}^{N} \gamma_i h_i''(x) \sum_{j=1}^{N} \gamma_j h_j''(x) dx$$

$$= \lambda \sum_{i=1}^{N} \sum_{j=1}^{N} \gamma_i \gamma_j \int_{-\infty}^{\infty} h_i''(x) h_j''(x) dx = \lambda \gamma^T G \gamma .$$

Thus, by differentiating $L(g)$ with respect to γ, as done to obtain the coefficients of ridge regression in Chap. 5, we find that the solution of

$$-X^T(y - X\gamma) + \lambda G\gamma = 0$$

is given by

$$\hat{\gamma} = (X^T X + \lambda G)^{-1} X^T y .$$

Because the proof of the following proposition is complicated, it is provided in the Appendix at the end of this chapter.

Proposition 22 *The elements $g_{i,j}$ defined in (7.6) are given by*

$$g_{i,j} = \frac{(x_{N-1} - x_{j-2})^2 \left(12x_{N-1} + 6x_{j-2} - 18x_{i-2}\right)}{+ 12(x_{N-1} - x_{i-2})(x_{N-1} - x_{j-2})(x_N - x_{N-1})}{(x_N - x_{i-2})(x_N - x_{j-2})}$$

for $x_i \le x_j$, where $g_{i,j} = 0$ for either $i \le 2$ or $j \le 2$.

For example, by means of the following procedure, we can obtain the matrix G from the knots $x_1 < \cdots < x_N$.

```
def G(x):
    n=len(x)
    g=np.zeros((n,n))
    for i in range(2,n):
        for j in range(i,n):
            g[i,j]=12*(x[n-1]-x[n-2])*(x[n-2]-x[j-2])*(x[n-1]-x[i-2])/(x[n
                -1]-x[i-2])/(x[n-1]-x[j-2])+(12*x[n-2]+6*x[j-2]-18*x[i-2])*(x[
                n-2]-x[j-2])**2/(x[n-1]-x[i-2])/(x[n-1]-x[j-2])
            g[j,i]=g[i,j]
    return g
```

Example 57 Computing the matrix G and $\hat{\gamma}$ for each λ, we draw the smoothing spline curve. We observe that the larger the λ is, the smoother the curve (Fig. 7.8). The procedure is implemented via the following code:

Fig. 7.8 In smoothing spline, we specify a parameter λ that expresses the smoothness instead of knots. For $\lambda = 40, 400, 1000$, we observe that the larger the λ is, the more difficult it is to fit the curve to the observed data

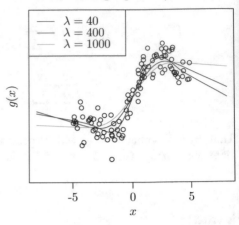

```
# generating data
n=100; a=-5; b=5
x=(b-a)*np.random.rand(n)+a  # uniform distribution (-5,5)
y=x-0.02*np.sin(x)-0.1*randn(n)
index=np.argsort(x); x=x[index]; y=y[index]

# compute x
X=np.zeros((n,n))
X[:,0]=1
for j in range(1,n):
    for i in range(n):
        X[i,j]=h(j,x[i],x)
GG=G(x)
lambda_set=[1,30,80]
col_set=["red","blue","green"]
plt.scatter(x,y,c="black",s=10)
plt.title("smoothing_splines_(n=100)")
plt.xlabel("x")
plt.ylabel("g(x)")
plt.tick_params(labelleft=False)

# smoothing splines when lambda=40, 400, 1000
for i in range(3):
    lam=lambda_set[i]
    gamma=np.linalg.inv(X.T@X+lam*GG)@X.T@y
    def g(u):
        S=gamma[0]
        for j in range(1,n):
            S=S+gamma[j]*h(j,u,x)
        return S
    u_seq=np.arange(-8,8,0.02)
    v_seq=[]
    for u in u_seq:
        v_seq.append(g(u))
    plt.plot(u_seq,v_seq,c=col_set[i],label="lambda={}".format(lambda_set[i
        ]))
plt.legend()
```

In ridge regression, we obtain a matrix of size $(p + 1) \times (p + 1)$. However, for the current problem, we must compute the inverse of a matrix of size $N \times N$, so we need an approximation because the computation is complex for large N.

However, if N is not large, the value of λ can be determined by cross-validation. Proposition 14 applies when matrix X is given by (7.4) with $K = N$. In addition, Proposition 15 applies when A is given by $X^T X + \lambda G$. Thus, the predictive error of CV in Proposition 14 is given by

$$CV[\lambda] := \sum_S \|(I - H_S[\lambda])^{-1} e_S\|^2 ,$$

where $H_S[\lambda] := X_S (X^T X + \lambda G)^{-1} X_S^T$. We construct the following procedure:

```python
def cv_ss_fast(X,y,lam,G,k):
    n=len(y)
    m=int(n/k)
    H=X@np.linalg.inv(X.T@X+lam*G)@X.T
    df=np.sum(np.diag(H))
    I=np.eye(n)
    e=(I-H)@y
    I=np.eye(m)
    S=0
    for j in range(k):
        test=np.arange(j*m,(j+1)*m)
        S=S+(np.linalg.inv(I-H[test,:][:,test])@e[test]).T@(np.linalg.inv(I-
            H[test,test])@e[test])
    return {'score':S/n,'df':df}
```

Note that if we set $\lambda = 0$, then the procedure is the same as `cv_fast` in Chap. 3.

How much the value of λ affects the estimation of γ depends on several conditions, and we cannot compare the λ values under different settings. Instead, we often use the *effective degrees of freedom*, the trace of the matrix $H[\lambda] := X(X^T X + \lambda G)^{-1} X^T$, rather than λ. The effective degrees of freedom express how well the fitness and simplicity are balanced (Fig. 7.9).

Fig. 7.9 The larger the λ is, the smaller the effective degrees of freedom. Even if the effective degrees of freedom are large, the predictive error of CV may increase

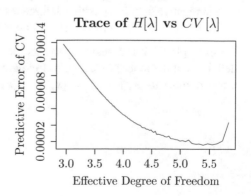

Trace of $H[\lambda]$ vs $CV[\lambda]$

Example 58 For sample size[1] $N = 100$, changing λ value from 1 to 50, we draw the graph of the effective degrees of freedom (the trace of $H[\lambda]$) and the predictive error of CV ($CV[\lambda]$). The execution is implemented via the following code.

```
# generating data
n=100; a=-5; b=5
x=(b-a)*np.random.rand(n)+a   # uniform distribution (-5,5)
y=x-0.02*np.sin(x)-0.1*randn(n)
index=np.argsort(x); x=x[index]; y=y[index]
#
X=np.zeros((n,n))
X[:,0]=1
for j in range(1,n):
    for i in range(n):
        X[i,j]=h(j,x[i],x)
GG=G(x)
# Calculations and plots of Effective Degree of Freedom and prediction
  errors
v=[]; w=[]
for lam in range(1,51,1):
    res=cv_ss_fast(X,y,lam,GG,n)
    v.append(res['df'])
    w.append(res['score'])
plt.plot(v,w)
plt.xlabel("Effective_Degree_of_Freedom")
plt.ylabel("prediction__errors_by_CV_")
plt.title("Effective_Degree_of_Freedom_and_prediction__errors_by_CV")
```

7.5 Local Regression

In this section, we consider the Nadaraya–Watson estimator and local linear regression.

Let $\mathcal{X}$ be a set. We call a function $k : \mathcal{X} \times \mathcal{X} \rightarrow \mathbb{R}$ a *kernel* (in a strict sense) if

1. for any $n \geq 1$ and $x_1, \ldots, x_n$, the matrix $K \in \mathcal{X}^{n \times n}$ with $K_{i,j} = k(x_i, x_j)$ is nonnegative definite (positive definiteness);
2. for any $x, y \in \mathcal{X}, k(x, y) = k(y, x)$ (symmetry).

For example, if $\mathcal{X}$ is a vector space, its inner product is a kernel. In fact, from the definition of the inner product $\langle \cdot, \cdot \rangle$: for elements x, y, and z of the vector space and a real number c, $\langle x, y + z \rangle = \langle x, y \rangle + \langle x, z \rangle$, $\langle cx, y \rangle = c \langle x, y \rangle$, $\langle x, x \rangle \geq 0$ for

---

[1]For $N > 100$, we could not compute the inverse matrix; errors occurred due to memory shortage.

arbitrary $a_1, \ldots, a_n \in \mathcal{X}$ and $c_1, \ldots, c_n \in \mathbb{R}$, we have

$$0 \le k \left(\sum_{i=1}^{n} c_i a_i, \sum_{j=1}^{n} c_j a_j \right) = \sum_i \sum_j c_i c_j k(a_i, a_j)$$

$$= [c_1, \ldots, c_n] \begin{bmatrix} k(a_1, a_1) & \cdots & k(a_1, a_n) \\ \vdots & \ddots & \vdots \\ k(a_n, a_1) & \cdots & k(a_n, a_n) \end{bmatrix} \begin{bmatrix} c_1 \\ \vdots \\ c_n \end{bmatrix}.$$

Kernels are used to express the similarity of two elements in set $\mathcal{X}$: the more similar the $x, y \in \mathcal{X}$ are, the larger the $k(x, y)$.

Even if $k : \mathcal{X} \times \mathcal{X} \to \mathbb{R}$ does not satisfy the positive definiteness, it can be used[2] if it accurately expresses the similarity.

Example 59 (Epanechnikov Kernel) The kernel $k : \mathcal{X} \times \mathcal{X} \to \mathbb{R}$ defined by

$$K_\lambda(x, y) = D \left(\frac{|x - y|}{\lambda} \right)$$

$$D(t) = \begin{cases} \frac{3}{4}(1 - t^2), & |t| \le 1 \\ 0, & \text{Otherwise} \end{cases}$$

does not satisfy positive definiteness. In fact, when $\lambda = 2, n = 3, x_1 = -1, x_2 = 0$, and $x_3 = 1$, the matrix with elements $K_\lambda(x_i, x_j)$ can be expressed as

$$\begin{bmatrix} K_\lambda(x_1, x_1) & K_\lambda(x_1, x_2) & K_\lambda(x_1, x_3) \\ K_\lambda(x_2, x_1) & K_\lambda(x_2, x_2) & K_\lambda(x_2, x_3) \\ K_\lambda(x_3, x_1) & K_\lambda(x_3, x_2) & K_\lambda(x_3, x_3) \end{bmatrix} = \begin{bmatrix} 3/4 & 9/16 & 0 \\ 9/16 & 3/4 & 9/16 \\ 0 & 9/16 & 3/4 \end{bmatrix}.$$

We see that the determinant is $3^3/2^6 - 3^5/2^{10} - 3^5/2^{10} = -3^3/2^9$. Since the determinant is equal to the product of the eigenvalues (Proposition 6), at least one of the three eigenvalues should be negative.

The Nadaraya–Watson estimator is constructed as

$$\hat{f}(x) = \frac{\sum_{i=1}^{N} K(x, x_i) y_i}{\sum_{j=1}^{N} K(x, x_j)}$$

from observed data $(x_1, y_1), \ldots, (x_N, y_N) \in \mathcal{X} \times \mathbb{R}$, where $\mathcal{X}$ is a set and $k : \mathcal{X} \times \mathcal{X} \to \mathbb{R}$ is a kernel. Then, given a new data point $x_* \in \mathcal{X}$, the estimator returns

---

[2]We call such a kernel a kernel in a broader sense.

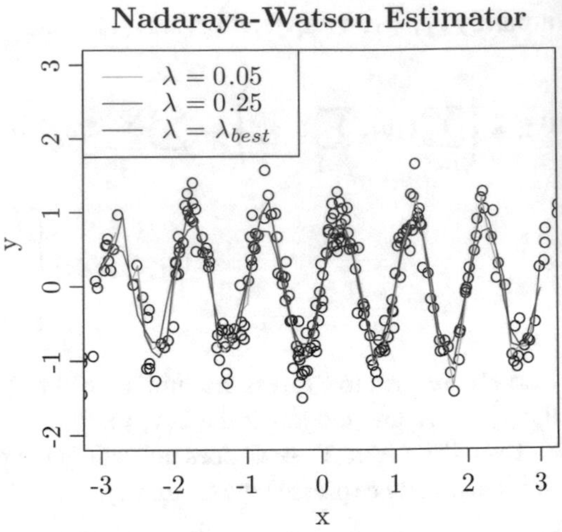

Fig. 7.10 We apply the
Epanechnikov kernel to the
Nadaraya–Watson estimator
and draw curves for
$\lambda = 0.05, 0.25$. Finally, we
compute the optimal λ and
draw the curve in the same
graph (Example 60)

$\hat{f}(x_*)$, which weights $y_1, \ldots, y_N$ according to the ratio

$$\frac{K(x_*, x_1)}{\sum_{j=1}^{N} K(x_*, x_j)}, \ldots, \frac{K(x_*, x_N)}{\sum_{j=1}^{N} K(x_*, x_j)}.$$

Since we assume that $k(u, v)$ expresses the similarity between $u, v \in \mathcal{X}$, the larger
the weight on y_i, the more similar x_* and x_i are.

Example 60 We apply the Epanechnikov kernel to the Nadaraya–Watson estimator. The Nadaraya–Watson estimator executes successfully even for kernels that do
not satisfy positive definiteness. For a given input $x_* \in \mathcal{X}$, the weights are only on
$y_i, i = 1, \ldots, N$, such that $x_i - \lambda \leq x_* \leq x_i + \lambda$. If the value of λ is small, the
prediction is made based on (x_i, y_i) such that x_i is within a small neighboring region
of x_*. We present the results obtained by executing the following code in Fig. 7.10.

```
n=250
x=2*randn(n)
y=np.sin(2*np.pi*x)+randn(n)/4
```

```
def D(t):
    return np.maximum(0.75*(1-t**2),0)
```

```
def K(x,y,lam):
    return D(np.abs(x-y)/lam)
```

```
def f(z,lam):
    S=0; T=0
    for i in range(n):
        S=S+K(x[i],z,lam)*y[i]
        T=T+K(x[i],z,lam)
    if T==0:
        return(0)
    else:
        return S/T
```

```
plt.scatter(x,y,c="black",s=10)
plt.xlim(-3,3)
plt.ylim(-2,3)
xx=np.arange(-3,3,0.1)
yy=[]
for zz in xx:
    yy.append(f(zz,0.05))
plt.plot(xx,yy,c="green",label="lambda=0.05")
yy=[]
for zz in xx:
    yy.append(f(zz,0.25))
plt.plot(xx,yy,c="blue",label="lambda=0.25")
# The curves of lam =0.05 , 0.25 were displayed.
m=int(n/10)
lambda_seq=np.arange(0.05,1,0.01)
SS_min=np.inf
for lam in lambda_seq:
    SS=0
    for k in range(10):
        test=list(range(k*m,(k+1)*m))
        train=list(set(range(n))-set(test))
        for j in test:
            u=0; v=0
            for i in train:
                kk=K(x[i],x[j],lam)
                u=u+kk*y[i]
                v=v+kk
            if v==0:
                d_min=np.inf
                for i in train:
                    d=np.abs(x[j]-x[i])
                    if d<d_min:
                        d_min=d
                        index=i
                z=y[index]
            else:
                z=u/v
            SS=SS+(y[j]-z)**2
    if SS<SS_min:
        SS_min=SS
        lambda_best=lam
yy=[]
for zz in xx:
    yy.append(f(zz,lambda_best))
plt.plot(xx,yy,c="red",label="lambda=lambda_best")
plt.title("Nadaraya-Watson estimator")
plt.legend()
```

We next consider local linear regression in which the coefficients are estimated for each local point (Fig. 7.11).

Fig. 7.11 We apply the
Epanechnikov kernel to draw
the graph of a local linear
regression curve
(Example 61): $p = 1$ and
$N = 30$

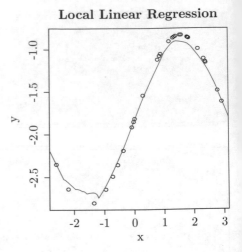

In standard linear regression, given observed data $(x_1, y_1), \ldots, (x_N, y_N) \in \mathbb{R}^p \times \mathbb{R}$, we obtain $\beta \in \mathbb{R}^{p+1}$ that minimizes

$$\sum_{i=1}^{N} (y_i - [1, x_i]\beta)^2 ,$$

where $x_i \in \mathbb{R}^p$ is a row vector. By contrast, in local linear regression, we obtain $\beta(x) \in \mathbb{R}^{p+1}$ that minimizes

$$\sum_{i=1}^{N} k(x, x_i)(y_i - [1, x_i]\beta(x))^2 \qquad (7.7)$$

for each $x \in \mathbb{R}^p$, where $k : \mathbb{R}^p \times \mathbb{R}^p \to \mathbb{R}$ is a kernel. Note that $\beta(x)$ depends on $x \in \mathbb{R}^p$, which is the main difference from standard local regression.

Equation (7.7) can be expressed as the matrix

$$(y - X\beta(x))^T \begin{bmatrix} k(x, x_1) & \cdots & 0 \\ \vdots & \ddots & \vdots \\ 0 & \cdots & k(x, x_N) \end{bmatrix} (y - X\beta(x)) , \qquad (7.8)$$

where the leftmost column of $X \in \mathbb{R}^{N \times (p+1)}$ is a column vector consisting of all ones. If we replace the diagonal matrix with the elements $k(x, x_1), \ldots, k(x, x_N)$ with W, then (7.8) is $(y - X\beta)^T W(y - X\beta)$, where W depends on x.

If we differentiate this equation with respect to β, we obtain

$$-2X^T W(y - X\beta(x)) .$$

Therefore, if we set this result to zero, we obtain $X^T W y = X^T W X \beta(x)$ and

$$\hat{\beta}(x) = (X^T W X)^{-1} X^T W y .$$

Example 61 We apply the Epanechnikov kernel with $p = 1$ to local linear regression and $x_1, \ldots, x_N$, and $y_1, \ldots, y_N$.

```python
def local(x,y,z=x):
    n=len(y)
    x=x.reshape(-1,1)
    X=np.insert(x,0,1,axis=1)
    yy=[]
    for u in z:
        w=np.zeros(n)
        for i in range(n):
            w[i]=K(x[i],u,lam=1)
        W=np.diag(w)
        beta_hat=np.linalg.inv(X.T@W@X)@X.T@W@y
        yy.append(beta_hat[0]+beta_hat[1]*u)
    return yy
```

```python
n=30
x=np.random.rand(n)*2*np.pi-np.pi
y=np.sin(x)+randn(1)
plt.scatter(x,y,s=15)
m=200
U=np.arange(-np.pi,np.pi,np.pi/m)
V=local(x,y,U)
plt.plot(U,V,c="red")
plt.title("
Local_linear_regression(p=1,N=30)")
```

7.6 Generalized Additive Models

If the number of basis functions is finite, we can obtain the coefficients as we did for linear regression.

Example 62 The basis of the polynomials of order $p = 4$ contains five functions $1, x, x^2, x^3$, and x^4, and the basis of the natural spline curves with $K = 5$ knots contains $1, x, h_3(x), h_4(x)$, and $h_5(x)$. However, if we mix them, we obtain eight linearly independent functions. We can estimate a function $f(x)$ that can be expressed by the sum of an order $p = 4$ polynomial and a $K = 5$-knot natural spline function as

$$\hat{f}(x) = \sum_{j=0}^{4} \hat{\beta}_j x^j + \sum_{j=5}^{7} \hat{\beta}_j h_{j-2}(x)$$

$\hat{\beta} = (X^T X)^{-1} X^T y = [\hat{\beta}_0, \ldots, \hat{\beta}_7]^T$ from observed data $(x_1, y_1), \ldots, (x_N, y_N)$, where

$$X = \begin{bmatrix} 1 & x_1 & x_1^2 & x_1^3 & x_1^4 & h_3(x_1) & h_4(x_1) & h_5(x_1) \\ 1 & x_2 & x_2^2 & x_2^3 & x_2^4 & h_3(x_2) & h_4(x_2) & h_5(x_2) \\ \vdots & \vdots & \vdots & \vdots & \vdots & \vdots & \vdots & \vdots \\ 1 & x_N & x_N^2 & x_N^3 & x_N^4 & h_3(x_N) & h_4(x_N) & h_5(x_N) \end{bmatrix}.$$

However, as for the smoothing spline curves with large sample size N, computing the inverse matrix is difficult. Moreover, in some cases, such as local regression, the curve cannot be expressed by a finite number of basis functions. In such cases, we often use a technique called *backfitting*. Suppose that we express a function $f(x)$ as the sum of functions $f_1(x), \ldots, f_p(x)$. We first set $f_1(x) = \cdots = f_p(x) = 0$, and for each $j = 1, \ldots, p$, we regress the residuals

$$r_j(x) := f(x) - \sum_{k \neq j} f_k(x)$$

on $f_j(x)$ and repeat the cycle until convergence.

Example 63 To divide the function into polynomial and local regression to understand the relation between covariates and responses, we implement the following procedure. We repeat the polynomial and local regressions in turn and divide $y \in \mathbb{R}^N$ into $y_1 + y_2 = y$. We present a graph that consists of the two elements in Fig. 7.12.

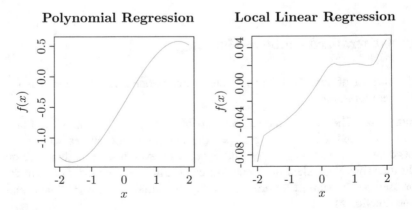

Fig. 7.12 We present the fitting via polynomial regression and local regression (Example 63)

```python
def poly(x,y,z=None):
    if z is None:
        z=x
    n=len(x)
    m=len(z)
    X=np.zeros((n,4))
    for i in range(n):
        X[i,0]=1; X[i,1]=x[i]; X[i,2]=x[i]**2; X[i,3]=x[i]**3
    beta_hat=np.linalg.inv(X.T@X)@X.T@y
    Z=np.zeros((m,4))
    for j in range(m):
        Z[j,0]=1; Z[j,1]=z[j]; Z[j,2]=z[j]**2; Z[j,3]=z[j]**3
    yy=Z@beta_hat
    return yy
```

```python
n=30
x=np.random.rand(n)*2*np.pi-np.pi
x=x.reshape(-1,1)
y=np.sin(x)+randn(n)
y_1=0; y_2=0
for k in range(10):
    y_1=poly(x,y-y_2)
    y_2=local(x,y-y_1,z=x)
z=np.arange(-2,2,0.1)
plt.plot(z,poly(x,y_1,z))
plt.title("polynomial_regression")
```

```python
plt.plot(z,local(x,y_2,z))
plt.title("Local_linear_regression")
```

Appendix: Proofs of Propositions

Proposition 20 *The function $f(x)$ has K cubic polynomials $h_1(x) = 1$, $h_2(x) = x$, $h_{j+2}(x) = d_j(x) - d_{K-1}(x)$, $j = 1, \ldots, K - 2$, as a basis, and if we define*

$$\gamma_1 := \beta_1, \ \gamma_2 := \beta_2, \ \gamma_3 := (\alpha_K - \alpha_1)\beta_3, \ \ldots, \ \gamma_K := (\alpha_K - \alpha_{K-2})\beta_K$$

for each $\beta_1, \ldots, \beta_K$, we can express f as $f(x) = \displaystyle\sum_{j=1}^{K} \gamma_j h_j(x)$, where we have

$$d_j(x) = \frac{(x - \alpha_j)_+^3 - (x - \alpha_K)_+^3}{\alpha_K - \alpha_j}, \ j = 1, \ldots, K - 1.$$

Proof First, the condition (7.3) $\beta_{K+1} = -\sum_{j=3}^{K} \dfrac{\alpha_K - \alpha_{j-2}}{\alpha_K - \alpha_{K-1}} \beta_j$ can be expressed as

$$\gamma_{K+1} = -\sum_{j=3}^{K} \gamma_j \tag{7.9}$$

with $\gamma_{K+1} := (\alpha_K - \alpha_{K-1})\beta_{K+1}$. $\square$

In the following, we show that $\gamma_1, \ldots, \gamma_K$ are coefficients when the basis consists of $h_1(x) = 1, h_2(x) = x, h_{j+2}(x) = d_j(x) - d_{K-1}(x), j = 1, \ldots, K - 2$, where

$$d_j(x) = \frac{(x - \alpha_j)_+^3 - (x - \alpha_K)_+^3}{\alpha_K - \alpha_j}, \quad j = 1, \ldots, K - 1$$

for each case of $x \le \alpha_K$ and $\alpha_K \le x$.

In fact, for $x \le \alpha_K$, using (7.9), we obtain

$$\sum_{j=3}^{K+1} \gamma_j \frac{(x - \alpha_{j-2})_+^3}{\alpha_K - \alpha_{j-2}} = \sum_{j=3}^{K} \gamma_j \frac{(x - \alpha_{j-2})_+^3}{\alpha_K - \alpha_{j-2}} - \sum_{j=3}^{K} \gamma_j \frac{(x - \alpha_{K-1})_+^3}{\alpha_K - \alpha_{K-1}}$$

$$= \sum_{j=3}^{K} \gamma_j \left\{ \frac{(x - \alpha_{j-2})_+^3}{\alpha_K - \alpha_{j-2}} - \frac{(x - \alpha_{K-1})_+^3}{\alpha_K - \alpha_{K-1}} \right\}$$

$$= \sum_{j=3}^{K} \gamma_j \{d_{j-2}(x) - d_{K-1}(x)\} ,$$

which means

$$f(x) = \beta_1 + \beta_2 x + \sum_{j=3}^{K+1} \beta_j (x - \alpha_{j-2})_+^3$$

$$= \gamma_1 + \gamma_2 x + \sum_{j=3}^{K+1} \gamma_j \frac{(x - \alpha_{j-2})_+^3}{\alpha_K - \alpha_{j-2}}$$

$$= \gamma_1 + \gamma_2 x + \sum_{j=3}^{K} \gamma_j (d_{j-2}(x) - d_{K-1}(x)) = \sum_{j=1}^{K} \gamma_j h_j(x) .$$

For $x \geq \alpha_K$, according to the definition, and $j = 1, \ldots, K - 2$, we have

$$h_{j+2}(x) = \frac{(x - \alpha_j)^3 - (x - \alpha_K)^3}{\alpha_K - \alpha_j} - \frac{(x - \alpha_{K-1})^3 - (x - \alpha_K)^3}{\alpha_K - \alpha_{K-1}}$$

$$= (x - \alpha_j)^2 + (x - \alpha_K)^2 + (x - \alpha_j)(x - \alpha_K) - (x - \alpha_K)^2$$

$$- (x - \alpha_{K-1})^2 - (x - \alpha_{K-1})(x - \alpha_K)$$

$$= (\alpha_{K-1} - \alpha_j)(2x - \alpha_j - \alpha_{K-1}) + (x - \alpha_K)(\alpha_{K-1} - \alpha_j) \quad (7.10)$$

$$= (\alpha_{K-1} - \alpha_j)(3x - \alpha_j - \alpha_{K-1} - \alpha_K), \quad (7.11)$$

where the second to last equality is obtained by factorization between the first and fourth terms and between the third and sixth terms. Therefore, if we substitute $x = \alpha_K$ into $f(x) = \sum_{j=1}^{K} \gamma_j h_j(x)$ and $f'(x) = \sum_{j=1}^{K} \gamma_j h'_j(x)$, we obtain

$$f(\alpha_K) = \gamma_1 + \gamma_2 \alpha_K + \sum_{j=3}^{K} \gamma_j (\alpha_{K-1} - \alpha_{j-2})(2\alpha_K - \alpha_{j-2} - \alpha_{K-1}) \quad (7.12)$$

and

$$f'(\alpha_K) = \gamma_2 + 3 \sum_{j=3}^{K} \gamma_j (\alpha_{K-1} - \alpha_{j-2}). \quad (7.13)$$

Thus, for $x \geq \alpha_K$, we have shown that $f(x) = \sum_{j=1}^{K} \gamma_j h_j(x)$ is such a line. On the other hand, using the function $f(x) = \gamma_1 + \gamma_2 x + \sum_{j=1}^{K+1} \gamma_j \frac{(x - \alpha_{j-2})_+^3}{\alpha_K - \alpha_{j-2}}$ for $x \leq \alpha_K$, to compute the value and its derivative at $x = \alpha_K$, from (7.9), we obtain

$$f(\alpha_K) = \gamma_1 + \gamma_2 \alpha_K + \sum_{j=3}^{K+1} \gamma_j \frac{(\alpha_K - \alpha_{j-2})^3}{\alpha_K - \alpha_{j-2}} = \gamma_1 + \gamma_2 \alpha_K + \sum_{j=3}^{K+1} \gamma_j (\alpha_K - \alpha_{j-2})^2$$

$$(7.14)$$

$$= \gamma_1 + \gamma_2 \alpha_K + \sum_{j=3}^{K} \gamma_j (\alpha_K - \alpha_{j-2})^2 - \sum_{j=3}^{K} \gamma_j (\alpha_K - \alpha_{K-1})^2$$

$$= \gamma_1 + \gamma_2 \alpha_K + \sum_{j=3}^{K} \gamma_j (\alpha_{K-1} - \alpha_{j-2})(2\alpha_K - \alpha_{j-2} - \alpha_{K-1}) \quad (7.15)$$

and

$$f'(\alpha_K) = \gamma_2 + 3 \sum_{j=3}^{K+1} \gamma_j \frac{(\alpha_K - \alpha_{j-2})^2}{\alpha_K - \alpha_{j-2}} = \gamma_2 + 3 \sum_{j=3}^{K+1} \gamma_j(\alpha_K - \alpha_{j-2}) \qquad (7.16)$$

$$= \gamma_2 + 3 \sum_{j=3}^{K} \gamma_j(\alpha_K - \alpha_{j-2}) - 3 \sum_{j=3}^{K} \gamma_j(\alpha_K - \alpha_{K-1})$$

$$= \gamma_2 + 3 \sum_{j=3}^{K} \gamma_j(\alpha_{K-1} - \alpha_{j-2}).$$

Since not only (7.12) and (7.15) but also (7.13) and (7.16) coincide, the proposition holds even for $x \geq \alpha_K$.

Proposition 21 (Green and Silverman, 1994) *The natural spline f with knots $x_1, \ldots, x_N$ minimizes $L(f)$.*

Proof Let $f(x)$ be an arbitrary function that minimizes (7.5), $g(x)$ be the natural spline with knots $x_1, \ldots, x_N$, and $r(x) := f(x) - g(x)$. Since the dimension of $g(x)$ is N, we can determine the coefficients $\gamma_1, \ldots, \gamma_N$ of the basis functions $h_1(x), \ldots, h_N(x)$ in $g(x) = \sum_{i=1}^{N} \gamma_i h_i(x)$ such that

$$g(x_1) = f(x_1), \ldots, g(x_N) = f(x_N).$$

In fact, we can solve the following linear equation:

$$\begin{bmatrix} h_1(x_1) & \cdots & h_N(x_1) \\ \vdots & \ddots & \vdots \\ h_1(x_N) & \cdots & h_N(x_N) \end{bmatrix} \begin{bmatrix} \gamma_1 \\ \vdots \\ \gamma_N \end{bmatrix} = \begin{bmatrix} f(x_1) \\ \vdots \\ f(x_N) \end{bmatrix}.$$

Then, note that we have $r(x_1) = \cdots = r_N(x_N) = 0$ and that $g(x)$ is a line and a cubic polynomial for $x \leq x_1$, $x_N \leq x$ and inside these values, respectively, which means that $g'''(x)$ is a constant γ_i for each interval $[x_i, x_{i+1}]$, specifically, $g''(x_1) = g''(x_N) = 0$. Thus, we have

$$\int_{x_1}^{x_N} g''(x)r''(x)dx = [g''(x)r'(x)]_{x_1}^{x_N} - \int_{x_1}^{x_N} g'''(x)r'(x)dx = -\sum_{i=1}^{N-1} \gamma_i[r(x)]_{x_i}^{x_{i+1}} = 0.$$

Hence, we have

$$\int_{-\infty}^{\infty} \{f''(x)\}^2 dx \geq \int_{x_1}^{x_N} \{g''(x) + r''(x)\}^2 dx$$

$$\geq \int_{x_1}^{x_N} \{g''(x)\}^2 dx + \int_{x_1}^{x_N} \{r''(x)\}^2 dx + 2\int_{x_1}^{x_N} g''(x)r''(x)dx$$

$$\geq \int_{x_1}^{x_N} \{g''(x)\}^2 dx \,,$$

which means that for each of the functions f that minimize $L(\cdot)$ in (7.5), there exists a natural function g such that

$$L(f) = \sum_{i=1}^{N}(y_i - f(x_i))^2 + \lambda \int_{-\infty}^{\infty} \{f''(x)\}^2 dx$$

$$\geq \sum_{i=1}^{N}(y_i - g(x_i))^2 + \lambda \int_{-\infty}^{\infty} \{g''(x)\}^2 dx = L(g) \,.$$

$\square$

Proposition 22 *The elements $g_{i,j}$ defined in (7.6) are given by*

$$g_{i,j} = \frac{(x_{N-1} - x_{j-2})^2 \left(12x_{N-1} + 6x_{j-2} - 18x_{i-2}\right)}{+12(x_{N-1} - x_{i-2})(x_{N-1} - x_{j-2})(x_N - x_{N-1})}}{(x_N - x_{i-2})(x_N - x_{j-2})} \,,$$

where $x_i \leq x_j$ and $g_{i,j} = 0$ for either $i \leq 2$ or $j \leq 2$.

Proof Without loss of generality, we may assume $x_i \leq x_j$. Then, we have

$$\int_{x_1}^{x_N} h_i''(x)h_j''(x)dx = \int_{\max(x_i,x_j)}^{x_N} h_i''(x)h_j''(x)dx$$

$$= \int_{x_j}^{x_{N-1}} h_i''(x)h_j''(x)dx + \int_{x_{N-1}}^{x_N} h_i''(x)h_j''(x)dx \,, \qquad (7.17)$$

where we have used $h_i''(x) = 0$ for $x \leq x_i$ and $h_j''(x) = 0$ for $x \leq x_j$. The right-hand side can be computed as follows. The second term is

$$
\int_{x_{N-1}}^{x_N} h_i''(x) h_j''(x) dx
$$

$$
= 36 \int_{x_{N-1}}^{x_N} \left(\frac{x - x_{i-2}}{x_N - x_{i-2}} - \frac{x - x_{N-1}}{x_N - x_{N-1}} \right) \left(\frac{x - x_{j-2}}{x_N - x_{j-2}} - \frac{x - x_{N-1}}{x_N - x_{N-1}} \right) dx
$$

$$
= 36 \frac{(x_{N-1} - x_{i-2})(x_{N-1} - x_{j-2})}{(x_N - x_{i-2})(x_N - x_{j-2})} \int_{x_{N-1}}^{x_N} \left(\frac{x - x_N}{x_N - x_{N-1}} \right)^2 dx
$$

$$
= 12 \frac{(x_{N-1} - x_{i-2})(x_{N-1} - x_{j-2})(x_N - x_{N-1})}{(x_N - x_{i-2})(x_N - x_{j-2})}, \tag{7.18}
$$

where the second equality is obtained via the following equations:

$$
(x - x_{i-2})(x_N - x_{N-1}) - (x - x_{N-1})(x_N - x_{i-2}) = (x - x_N)(x_{N-1} - x_{i-2})
$$

$$
(x - x_{j-2})(x_N - x_{N-1}) - (x - x_{N-1})(x_N - x_{j-2}) = (x - x_N)(x_{N-1} - x_{j-2}) .
$$

For the first term of (7.17), we have

$$
\int_{x_{j-2}}^{x_{N-1}} h_i''(x) h_j''(x) dx = 36 \int_{x_{j-2}}^{x_{N-1}} \frac{x - x_{i-2}}{x_N - x_{i-2}} \cdot \frac{x - x_{j-2}}{x_N - x_{j-2}} dx
$$

$$
= 36 \frac{x_{N-1} - x_{j-2}}{(x_N - x_{i-2})(x_N - x_{j-2})}
$$

$$
\times \left\{ \frac{1}{3}(x_{N-1}^2 + x_{N-1}x_{j-2} + x_{j-2}^2) - \frac{1}{2}(x_{N-1} + x_{j-2})(x_{i-2} + x_{j-2}) + x_{i-2}x_{j-2} \right\}
$$

$$
= 36 \frac{x_{N-1} - x_{j-2}}{(x_N - x_{i-2})(x_N - x_{j-2})} \left\{ \frac{1}{3}x_{N-1}^2 - \frac{1}{6}x_{N-1}x_{j-2} - \frac{1}{6}x_{j-2}^2 - \frac{1}{2}x_{i-2}(x_{N-1} - x_{j-2}) \right\}
$$

$$
= \frac{(x_{N-1} - x_{j-2})^2}{(x_N - x_{i-2})(x_N - x_{j-2})} \left(12x_{N-1} + 6x_{j-2} - 18x_{i-2} \right), \tag{7.19}
$$

where to obtain the last equality in (7.19), we used

$$
\frac{1}{3}x_{N-1}^2 - \frac{1}{6}(x_{j-2} + 3x_{i-2})x_{N-1} - \frac{1}{6}x_{j-2}(x_{j-2} - 3x_{i-2})
$$

$$
= (x_{N-1} - x_{j-2})(\frac{1}{3}x_{N-1} + \frac{1}{6}x_{j-2} - \frac{1}{2}x_{i-2}).
$$

$\square$

Exercises 57–68

57. For each of the following two quantities, find a condition under which the $\beta_0, \beta_1, \ldots, \beta_p$ that minimize it are unique given data $(x_1, y_1), \ldots, (x_N, y_N) \in \mathbb{R} \times \mathbb{R}$ and its solution:

(a) $\displaystyle\sum_{i=1}^{N} \left(y_i - \sum_{j=0}^{p} \beta_j x_i^j \right)^2$

(b) $\displaystyle\sum_{i=1}^{N} \left(y_i - \sum_{j=0}^{p} \beta_j f_j(x_i) \right)^2$, $f_0(x) = 1, x \in \mathbb{R}, f_j : \mathbb{R} \to \mathbb{R}, j = 1, \ldots, p.$

58. For $K \geq 1$ and $-\infty = \alpha_0 < \alpha_1 < \cdots < \alpha_K < \alpha_{K+1} = \infty$, we define a cubic polynomial $f_i(x)$ for $\alpha_i \leq x \leq \alpha_{i+1}, i = 0, 1, \ldots, K$, and assume that $f_i, i = 0, 1, \ldots, K$, satisfy $f_{i-1}^{(j)}(\alpha_i) = f_i^{(j)}(\alpha_i), j = 0, 1, 2, i = 1, \ldots, K$, where $f^{(0)}(\alpha), f^{(1)}(\alpha)$, and $f^{(2)}(\alpha)$ denote the value, the first, and the second derivatives of f at $x = \alpha$.

(a) Show that there exists γ_i such that $f_i(x) = f_{i-1}(x) + \gamma_i(x - \alpha_i)^3$.
(b) Consider a piecewise cubic polynomial $f(x) = f_i(x)$ for $\alpha_i \leq x \leq \alpha_{i+1}$ $i = 0, 1, \ldots, K$ (spline curve). Show that there exist $\beta_1, \beta_2, \ldots, \beta_{K+4}$ such that

$$f(x) = \beta_1 + \beta_2 x + \beta_3 x^2 + \beta_4 x^3 + \sum_{i=1}^{K} \beta_{i+4}(x - \alpha_i)_+^3 ,$$

where $(x - \alpha_i)_+$ denotes the function that takes $x - \alpha_i$ and zero for $x > \alpha_i$ and $x \leq \alpha_i$, respectively.

59. We generate artificial data and execute spline regression for $K = 5, 7, 9$ knots. Define the following function f and draw spline curves.

```
n=100
x=randn(n)*2*np.pi
y=np.sin(x)+0.2*randn(n)
col_set=["red","green","blue"]
K_set=[5,7,9]
plt.scatter(x,y,c="black",s=10)
plt.xlim(-5,5)
for k in range(3):
    K=K_set[k]
    knots=np.linspace(-2*np.pi,2*np.pi,K)
    X=np.zeros((n,K+4))
    for i in range(n):
        X[i,0]=1
        X[i,1]=x[i]
        X[i,2]=x[i]**2
        X[i,3]=x[i]**3
        for j in range(K):
            X[i,j+4]=np.maximum((x[i]-knots[j])**3,0)
```

```
    beta=np.linalg.inv(X.T@X)@X.T@y
    # some blanks (definition of function f)#
    u_seq=np.arange(-5,5,0.02)
    v_seq=[]
    for u in u_seq:
        v_seq.append(f(u))
    plt.plot(u_seq,v_seq,c=col_set[k],label="K={}".format(K))
plt.legend()
```

60. For $K \geq 2$, we define the following cubic spline curve g (natural spline): it is a line for $x \leq \alpha_1$ and $\alpha_K \leq x$ and a cubic polynomial for $\alpha_i \leq x \leq \alpha_{i+1}$, $i = 1, \ldots, K-1$, where the values and the first and second derivatives coincide on both sides of the K knots $\alpha_1, \ldots, \alpha_K$.

(a) Show that $\gamma_{K+1} = -\sum_{j=3}^{K} \gamma_j$ when

$$g(x) = \gamma_1 + \gamma_2 x + \gamma_3 \frac{(x-\alpha_1)^3}{\alpha_K - \alpha_1} + \cdots + \gamma_K \frac{(x-\alpha_{K-2})^3}{\alpha_K - \alpha_{K-2}} + \gamma_{K+1} \frac{(x-\alpha_{K-1})^3}{\alpha_K - \alpha_{K-1}}$$

for $\alpha_{K-1} \leq x \leq \alpha_K$. Hint: Derive the result from $g''(\alpha_K) = 0$.

(b) $g(x)$ can be written as $\sum_{i=1}^{K} \gamma_i h_i(x)$ with $\gamma_1, \ldots, \gamma_K \in \mathbb{R}$ and the functions $h_1(x) = 1$, $h_2(x) = x$, $h_{j+2}(x) = d_j(x) - d_{K-1}(x)$, $j = 1, \ldots, K-2$, where

$$d_j(x) = \frac{(x-\alpha_j)_+^3 - (x-\alpha_K)_+^3}{\alpha_K - \alpha_j}, \quad j = 1, \ldots, K-1.$$

Show that

$$h_{j+2}(x) = (\alpha_{K-1} - \alpha_j)(3x - \alpha_j - \alpha_{K-1} - \alpha_K), \quad j = 1, \ldots, K-2$$

for each $\alpha_K \leq x$.

(c) Show that $g(x)$ is a linear function of x for $x \leq \alpha_1$ and for $\alpha_K \leq x$.

61. We compare the ordinary and natural spline functions. Define the functions $h_1, \ldots, h_K, d_1, \ldots, d_{K-1}$, and g, and execute the below:

```
def d(j,x,knots):
    # some blanks (definition of function d)#
```

```
def h(j,x,knots):
    # some blanks (definition of function h)#
```

```
n=100
x=randn(n)*2*np.pi
y=np.sin(x)+0.2*randn(n)
K=11
knots=np.linspace(-5,5,K)
X=np.zeros((n,K+4))
for i in range(n):
    X[i,0]=1
    X[i,1]=x[i]
    X[i,2]=x[i]**2
    X[i,3]=x[i]**3
    for j in range(K):
        X[i,j+4]=np.maximum((x[i]-knots[j])**3,0)
beta=np.linalg.inv(X.T@X)@X.T@y
```

```
def f(x):
    S=beta[0]+beta[1]*x+beta[2]*x**2+beta[3]*x**3
    for j in range(K):
        S=S+beta[j+4]*np.maximum((x-knots[j])**3,0)
    return S
```

```
X=np.zeros((n,K))
X[:,0]=1
for j in range(1,K):
    for i in range(n):
        X[i,j]=h(j,x[i],knots)
gamma=np.linalg.inv(X.T@X)@X.T@y
```

```
def g(x):
    # some blanks (definition of function g)#
```

```
u_seq=np.arange(-6,6,0.02)
v_seq=[]; w_seq=[]
for u in u_seq:
    v_seq.append(f(u))
    w_seq.append(g(u))
plt.scatter(x,y,c="black",s=10)
plt.xlim(-6,6)
plt.xlabel("x")
plt.ylabel("f(x),g(x)")
plt.tick_params(labelleft=False)
plt.plot(u_seq,v_seq,c="blue",label="spline_")
plt.plot(u_seq,w_seq,c="red",label="natural_spline")
plt.vlines(x=[-5,5],ymin=-1.5,ymax=1.5,linewidth=1)
plt.vlines(x=knots,ymin=-1.5,ymax=1.5,linewidth=0.5,linestyle="dashed")
plt.legend()
```

Hint: The functions h and d need to compute the size K of the knots. Inside the function g, *knots* may be global.

62. We wish to prove that for an arbitrary $\lambda \geq 0$, there exists $f : \mathbb{R} \to \mathbb{R}$ that minimizes

$$RSS(f, \lambda) := \sum_{i=1}^{N} (y_i - f(x_i))^2 + \lambda \int_{-\infty}^{\infty} \{f''(t)\}^2 dt, \qquad (7.20)$$

given data $(x_1, y_1), \ldots, (x_N, y_N) \in \mathbb{R} \times \mathbb{R}$ among the natural spline function g with knots $x_1 < \cdots < x_N$ (smoothing spline function).

(a) Show that there exist $\gamma_1, \ldots, \gamma_{N-1} \in \mathbb{R}$ such that

$$\int_{x_1}^{x_N} g''(x) r''(x) dx = -\sum_{i=1}^{N-1} \gamma_i \{r(x_{i+1}) - r(x_i)\}.$$

Hint: Use the facts that $g''(x_1) = g''(x_N) = 0$ and that the third derivative of g is constant for $x_i \leq x \leq x_{i+1}$.

(b) Show that if the function $h : \mathbb{R} \to \mathbb{R}$ satisfies

$$\int_{x_1}^{x_N} g''(x) r''(x) dx = 0, \qquad (7.21)$$

then for any $f(x) = g(x) + h(x)$, we have

$$\int_{-\infty}^{\infty} \{g''(x)\}^2 dx \leq \int_{-\infty}^{\infty} \{f''(x)\}^2 dx . \qquad (7.22)$$

Hint: For $x \leq x_1$ and $x_N \leq x$, $g(x)$ is a linear function and $g''(x) = 0$. Moreover, (7.21) implies

$$\int_{x_1}^{x_N} \{g''(x) + r''(x)\}^2 dx = \int_{x_1}^{x_N} \{g''(x)\}^2 dx + \int_{x_1}^{x_N} \{r''(x)\}^2 dx .$$

(c) A natural spline curve g is contained among the set of functions $f : \mathbb{R} \to \mathbb{R}$ that minimize (7.20). Hint: Show that if $RSS(f, \lambda)$ is the minimum value, $r(x_i) = 0$, $i = 1, \ldots, N$, implies (7.21) for the natural spline g such that $g(x_i) = f(x_i)$, $i = 1, \ldots, N$.

63. It is known that $g_{i,j} := \int_{-\infty}^{\infty} h_i''(x) h_j''(x) dx$ is given by

$$\frac{(x_{N-1} - x_{j-2})^2 (12x_{N-1} - 18x_{i-2} + 6x_{j-2})}{+ 12(x_{N-1} - x_{i-2})(x_{N-1} - x_{j-2})(x_N - x_{N-1})}{(x_N - x_{i-2})(x_N - x_{j-2})},$$

where $h_1, \ldots, h_K$ is the natural spline basis with the knots $x_1 < \cdots < x_K$ and $g_{i,j} = 0$ for either $i \leq 2$ or $j \leq 2$. Write a Python function G that outputs matrix G with elements $g_{i,j}$ from the K knots $x \in \mathbb{R}^K$.

64. We assume that there exist $\gamma_1, \ldots, \gamma_N \in \mathbb{R}$ such that $g(x) = \sum_{j=1}^{N} g_j(x)\gamma_j$ and

$g''(x) = \sum_{j=1}^{N} g_j''(x)\gamma_j$ for a smoothing spline function g with knots $x_1 < \cdots <$

x_N, where g_j, $j = 1, \ldots, N$ are cubic polynomials. Show that the coefficients $\gamma = [\gamma_1, \ldots, \gamma_N]^T \in \mathbb{R}^N$ can be expressed by $\gamma = (G^T G + \lambda G'')^{-1} G^T y$ with $G = (g_j(x_i)) \in \mathbb{R}^{N \times N}$ and $G'' = \left(\int_{-\infty}^{\infty} g_j''(x) g_k''(x) dx \right) \in \mathbb{R}^{N \times N}$. Moreover,

we wish to draw the smoothing spline curve to compute $\hat{\gamma}$ for each λ. Fill in the blanks and execute the procedure.

```
# generating data
n=100; a=-5; b=5
x=(b-a)*np.random.rand(n)+a    # uniform distribution (-5,5)
y=x-0.02*np.sin(x)-0.1*randn(n)
index=np.argsort(x); x=x[index]; y=y[index]
```

```
X=np.zeros((n,n))
X[:,0]=1
for j in range(1,n):
    for i in range(n):
        X[i,j]=h(j,x[i],x)
GG=G(x)
lambda_set=[10,30,80]
col_set=["red","blue","green"]
plt.scatter(x,y,c="black",s=10)
plt.title("smoothing_spline(n=100)")
plt.xlabel("x")
plt.ylabel("g(x)")
plt.tick_params(labelleft=False)
for i in range(3):
    lam=lambda_set[i]
    gamma=# blank #
    def g(u):
        S=gamma[0]
        for j in range(1,n):
            S=S+gamma[j]*h(j,u,x)
        return S
    u_seq=np.arange(-8,8,0.02)
    v_seq=[]
    for u in u_seq:
        v_seq.append(g(u))
    plt.plot(u_seq,v_seq,c=col_set[i],label="lambda={}".format(
        lambda_set[i]))
plt.legend()
```

65. It is difficult to evaluate how much the value of λ affects the estimation of γ because λ varies and depends on the settings. To this end, we often use the effective degrees of freedom, the trace of $H[\lambda] := X(X^T X + \lambda G)^{-1} X^T$, instead of λ to evaluate the balance between fitness and simplicity. For $N = 100$

and λ ranging from 1 to 50, we draw the graph of the effective degrees of freedom (the trace of $H[\lambda]$) and predictive error $(CV[\lambda])$ of CV. Fill in the blanks and execute the procedure.

```
def cv_ss_fast(X,y,lam,G,k):
    n=len(y)
    m=int(n/k)
    H=X@np.linalg.inv(X.T@X+lam*G)@X.T
    df=# blank(1) #
    I=np.eye(n)
    e=(I-H)@y
    I=np.eye(m)
    S=0
    for j in range(k):
        test=np.arange(j*m,(j+1)*m)
        S=S+(np.linalg.inv(I-H[test,:][:,test])@e[test]).T@(np.linalg.
            inv(I-H[test,test])@e[test])
    return  {'score':S/n,'df':df}
```

```
# generating data
n=100; a=-5; b=5
x=(b-a)*np.random.rand(n)+a  # (-5,5)
y=x-0.02*np.sin(x)-0.1*randn(n)
index=np.argsort(x); x=x[index]; y=y[index]

# calculate X
X=np.zeros((n,n))
X[:,0]=1
for j in range(1,n):
    for i in range(n):
        X[i,j]=h(j,x[i],x)
GG=G(x)
# Calculations and plots of Effective Degree of Freedom and prediction
    errors
v=[]; w=[]
for lam in range(1,51,1):
    res=cv_ss_fast(# blank(2) #,n)
    v.append(res['df'])
    w.append(res['score'])
plt.plot(v,w)
plt.xlabel("Effective_Degree_of_Freedom")
plt.ylabel("prediction__errors_by_CV_")
plt.title("Effective_Degree_of_Freedom_and_prediction__errors_by_CV__")
```

66. Using the Nadaraya–Watson estimator

$$\hat{f}(x) = \frac{\sum_{i=1}^{N} K_\lambda(x, x_i) y_i}{\sum_{i=1}^{N} K_\lambda(x, x_i)}$$

with $\lambda > 0$ and the following kernel

$$K_\lambda(x, y) = D\left(\frac{|x - y|}{\lambda}\right)$$

$$D(t) = \begin{cases} \dfrac{3}{4}(1 - t^2), & |t| \leq 1 \\ 0, & \text{Otherwise}, \end{cases}$$

we draw a curve that fits $n = 250$ data. Fill in the blanks and execute the procedure. When λ is small, how does the curve change?

```
n=250
x=2*randn(n)
y=np.sin(2*np.pi*x)+randn(n)/4
```

```
def D(t):
    # some blanks (definition of function D)#
```

```
def K(x,y,lam):
    # some blanks (definition of function K)#
```

```
def f(z,lam):
    S=0; T=0
    for i in range(n):
        S=S+K(x[i],z,lam)*y[i]
        T=T+K(x[i],z,lam)
    if T==0:
        return(0)
    else:
        return S/T
```

```
plt.scatter(x,y,c="black",s=10)
plt.xlim(-3,3)
plt.ylim(-2,3)
xx=np.arange(-3,3,0.1)
yy=[]
for zz in xx:
    yy.append(f(zz,0.05))
plt.plot(xx,yy,c="green",label="lambda=0.05")
yy=[]
for zz in xx:
    yy.append(f(zz,0.25))
plt.plot(xx,yy,c="blue",label="E=0.25")
# # The curves of lam =0.05 , 0.25 were displayed.
m=int(n/10)
lambda_seq=np.arange(0.05,1,0.01)
SS_min=np.inf
for lam in lambda_seq:
    SS=0
    for k in range(10):
        test=list(range(k*m,(k+1)*m))
        train=list(set(range(n))-set(test))
        for j in test:
            u=0; v=0
            for i in train:
                kk=K(x[i],x[j],lam)
                u=u+kk*y[i]
                v=v+kk
            if v==0:
                d_min=np.inf
                for i in train:
                    d=np.abs(x[j]-x[i])
```

```
               if d<d_min:
                   d_min=d
                   index=i
           z=y[index]
       else:
           z=u/v
       SS=SS+(y[j]-z)**2
   if SS<SS_min:
       SS_min=SS
       lambda_best=lam
yy=[]
for zz in xx:
   yy.append(f(zz,lambda_best))
plt.plot(xx,yy,c="red",label="lam=lambda_best")
plt.title("Nadaraya-Watson_estimator")
plt.legend()
```

67. Let K be a kernel. We can obtain the predictive value $[1, x]\beta(x)$ for each $x \in \mathbb{R}^p$ using the $\beta(x) \in \mathbb{R}^{p+1}$ that minimizes

$$\sum_{i=1}^{N} K(x, x_i)(y_i - [1, x_i]\beta(x))^2$$

(local regression).

(a) When we write $\beta(x) = (X^T W(x)X)^{-1}X^T W(x)y$, what is the matrix W?
(b) Using the same kernel as we used in Problem 66 with $p = 1$, we applied $x_1, \ldots, x_N, y_1, \ldots, y_N$ to local regression. Fill in the blanks and execute the procedure.

```
def local(x,y,z=x):
    n=len(y)
    x=x.reshape(-1,1)
    X=np.insert(x,0,1,axis=1)
    yy=[]
    for u in z:
        w=np.zeros(n)
        for i in range(n):
            w[i]=K(x[i],u,lam=1)
        W=# blank(1) #
        beta_hat=# blank(2) #
        yy.append(beta_hat[0]+beta_hat[1]*u)
    return yy
```

```
n=30
x=np.random.rand(n)*2*np.pi-np.pi
y=np.sin(x)+randn(n)
plt.scatter(x,y,s=15)
m=200
U=np.arange(-np.pi,np.pi,np.pi/m)
V=local(x,y,U)
plt.plot(U,V,c="red")
plt.title("Local_linear_regression_(p=1,N=30)")
```

68. If the number of base functions is finite, the coefficient can be obtained via least squares in the same manner as linear regression. However, when the number

of bases is large, such as for the smoothing spline, it is difficult to find the inverse matrix. Moreover, for example, local regression cannot be expressed by a finite number of bases. In such cases, a method called backfitting is often applied. To decompose the function into the sum of polynomial regression and local regression, we constructed the following procedure. Fill in the blanks and execute the process.

```python
def poly(x,y,z=x):
    n=len(x)
    m=len(z)
    X=np.zeros((n,4))
    for i in range(n):
        X[i,0]=1; X[i,1]=x[i]; X[i,2]=x[i]**2; X[i,3]=x[i]**3
    beta_hat=np.linalg.inv(X.T@X)@X.T@y
    Z=np.zeros((m,4))
    for j in range(m):
        Z[j,0]=1; Z[j,1]=z[j]; Z[j,2]=z[j]**2; Z[j,3]=z[j]**3
    yy=# blank(1) #
    return yy
```

```python
n=30
x=np.random.rand(n)*2*np.pi-np.pi
x=x.reshape(-1,1)
y=np.sin(x)+randn(n)
y_1=0; y_2=0
for k in range(10):
    y_1=poly(x,y-y_2)
    y_2=# Blank #
z=np.arange(-2,2,0.1)
plt.plot(z,poly(x,y_1,z))
plt.title("_polynomial_regression")
```

```python
plt.plot(z,local(x,y_2,z))
plt.title("Local_Linear_Regression")
```

Chapter 8
Decision Trees

Abstract In this chapter, we construct decision trees by estimating the relationship between the covariates and the response from observed data. Starting from the root, each vertex traces to either the left or right at each branch, depending on whether a condition w.r.t. the covariates is met, and finally reaches a terminal node to obtain the response. Compared with the methods we have considered thus far, since it is expressed as a simple structure, the estimation accuracy of a decision tree is poor, but since it is expressed visually, it is easy to understand the relationship between the covariates and the response. Decision trees are often used to understand relationships rather than to predict the future, and decision trees can be used for regression and classification. The decision tree has the problem that the estimated tree shapes differ greatly even if observation data that follow the same distribution are used. Therefore, similar to the bootstrap discussed in Chap. 4, by sampling data of the same size from the original data multiple times, we reduce the variation in the obtained decision tree and this improvement can be considered. Finally, we introduce a method (boosting) that produces many small decision trees in the same way as the backfitting method learned in Chap. 7 to make highly accurate predictions.

8.1 Decision Trees for Regression

We wish to illustrate the relationship between the covariates (p variables) and the response by means of the observed data $(x_1, y_1), \ldots, (x_N, y_N) \in \mathbb{R}^p \times \mathbb{R}$. To this end, we consider constructing a decision tree. A decision tree consists of vertices and branches. Vertices that branch left and right are called branch nodes or interior nodes, and vertices that do not branch are called terminal nodes. Of the two adjacent vertices on a branch, the one closest to the terminal point is called the child, and the other vertex is called the parent. Furthermore, vertices that do not have parents are called roots (Fig. 8.1). When we construct a decision tree, each $x \in \mathbb{R}^p$ belongs to one of the regions $R_1, \ldots, R_m$ that correspond to terminal nodes. Then, for both regression and classification decision trees, two values in the same region should

Fig. 8.1 Decision trees.
Each vertex is either an inner
or terminal node. In this book,
we direct the edges from the
parents to their children

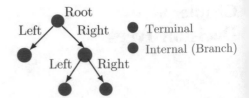

output the same response. Specifically, when the joint probability density function
is $f_{XY}(x, y)$, we construct a rule such that

$$x_i \in R_j \implies \hat{y}_i = \bar{y}_j \tag{8.1}$$

with

$$\bar{y}_j := E[Y|R_j] = \frac{\int_{-\infty}^{\infty} \int_{R_j} y f_{XY}(x, y) dx dy}{\int_{-\infty}^{\infty} \int_{R_j} f_{XY}(x, y) dx dy} \tag{8.2}$$

by obtaining $m \geq 1$ and the regions $R_1, \ldots, R_m$ that minimize

$$\int_{-\infty}^{\infty} \sum_{j=1}^{m} \int_{R_j} (y - \bar{y}_j)^2 f_{XY}(x, y) dx dy . \tag{8.3}$$

However, we must consider some problems to actually construct a decision tree
from the samples. First, the simultaneous density function f_{XY} is unknown and
should be estimated from the samples. To this end, if the size of the region R_j is n_j,
then one might replace (8.1), (8.2), and (8.3) by

$$x \in R_j \implies \hat{y}_i = \bar{y}_j ,$$

$$\bar{y}_j := \frac{1}{n_j} \sum_{i:x_i \in R_j} y_i ,$$

and

$$\sum_{j=1}^{m} \sum_{i:x_i \in R_j} (y_i - \bar{y}_j)^2 , \tag{8.4}$$

where $\displaystyle\sum_{i:x_i \in R_j} \cdot$ sums over i such that $x_i \in R_j$.

However, if we minimize (8.4) to obtain the regions, similar to the RSS value in
linear regression, the greater the value of m is, the smaller the value of (8.4). We see
that (8.4) always has the minimum value (zero) for $m = N$ (one sample per region).

Overfitting occurs because we use the same samples for testing and training. In fact, the testing using data other than the training data will be better if the size of each region R_j is greater than one, although the performance will be worse if the region size is too large. Thus, we may consider either separating the training and test data or applying cross-validation (CV).

A typical way to explicitly avoid overfitting is to obtain $m \geq 1$ and $R_1, \ldots, R_m$ that minimize

$$\sum_{j=1}^{m} \sum_{i:x_i \in R_j} (y_i - \bar{y}_j)^2 + \alpha m , \tag{8.5}$$

where the value of $\alpha > 0$ can be obtained via CV. For example, for each α, 90% of the data are used for training to obtain a decision tree that minimizes (8.5). The remaining data (10%) are used for testing to evaluate the performance of the decision tree. By using a different 10% as test data 10 times and taking the arithmetic average, we obtain the performance of a specific $\alpha > 0$. We obtain such an evaluation for all α values and choose the best decision tree (Fig. 8.2).

In addition, there are degrees of freedom, including how many variables are used at each branch node and how many branches are used. For example, if the positive and negative values of the linear sum of multiple variables are included, the number of combinations increases. An optimal number of branches may exist for each branch node. Furthermore, if the variables used for branching at each branch node are decided from the top down, the optimum decision tree cannot be obtained. It is necessary to look at all the vertices of the decision tree simultaneously and select the optimal combination of variables. Therefore, the optimal solution cannot be obtained. In this chapter, we choose one variable (X_j) and one sample ($x_{i,j}$ is the threshold) for each branching from the root to each terminal node in a greedy and top-down manner. Often, the threshold value is set to be equal to or greater than one of the samples in the node.

Suppose that a branch node contains samples x_i with $i \in S$, and by the sample size S, we mean the cardinality of S, where S is a subset of $\{1, 2, \cdots, N\}$. In the following, we divide a subset $\{x_k | k \in S\}$ of $\{x_1, \ldots, x_N\}$ into $\{x_k | x_k < x_{i,j}, k \in S\}$ and $\{x_k | x_k \geq x_{i,j}, k \in S\}$ by specifying $i \in S$ and $j = 1, \cdots p$. Although this approach does not take overfitting into consideration, we may select i, j that minimize

$$\sum_{k:x_{k,j} < x_{i,j}} (y_i - \bar{y}_{i,j}^L)^2 + \sum_{k:x_{k,j} \geq x_{i,j}} (y_i - \bar{y}_{i,j}^R)^2 ,$$

where n_L and n_R are the numbers of k such that $x_{k,j} < x_{i,j}$ and $x_{k,j} \geq x_{i,j}$, respectively, and $\bar{y}_{i,j}^L$ and $\bar{y}_{i,j}^R$ are $(1/n_L) \sum_{k:x_{k,j} < x_{i,j}} y_k$ and $(1/n_R) \sum_{k:x_{k,j} \geq x_{i,j}} y_k$, respectively.

We construct the following procedure for regression, where we apply a loss measure such as sq_loss to the argument f of function branch. Specifically, the

$\alpha = 0$ $\alpha = 0.1$

$\alpha = 0.5$ $\alpha = 2$

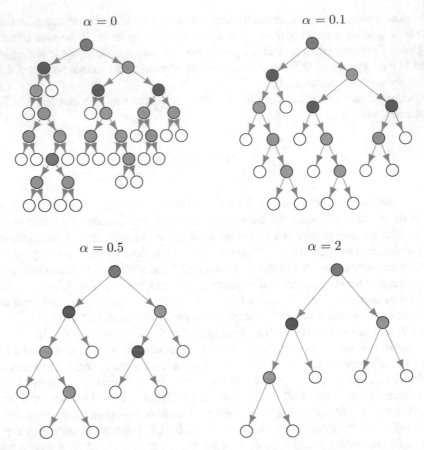

Fig. 8.2 For parameters $\alpha = 0, 0.1, 0.5, 2$, we generate decision trees that minimize (8.5). If the colors of the inner nodes are the same, the variables used for branching are the same. If attribute j is zero, the node is terminal. The larger the α is, the smaller the depth of the tree because pruning is performed in an earlier stage

procedure chooses the i, j that maximize the decrease before and after branching, assuming that each $k \in S$ proceeds to the left and right depending on $x_{k,j} < x_{i,j}$ and $x_{k,j} \geq x_{i,j}$.

```
def sq_loss(y):
    if len(y)==0:
        return 0
    else:
        y_bar=np.mean(y)
        return np.linalg.norm(y-y_bar)**2
```

```
def branch(x,y,S,rf=0):
    if rf==0:
        m=x.shape[1]
```

```
if x.shape[0]==0:
    return([0,0,0,0,0,0,0])
best_score=np.inf
for j in range(x.shape[1]):
    for i in S:
        left=[]; right=[]
        for k in S:
            if x[k,j]<x[i,j]:
                left.append(k)
            else:
                right.append(k)
        left_score=f(y[left]); right_score=f(y[right])
        score=left_score+right_score
        if score<best_score:
            best_score=score
            i_1=i; j_1=j
            left_1=left; right_1=right
            left_score_1=left_score; right_score_1=right_score
return [i_1,j_1,left_1,right_1,best_score,left_score_1,right_score_1]
```

In the above procedure, the samples x_i, $i \in S$ in the node are the candidate thresholds. However, if the sample size N is large, we may reduce the candidates to avoid enormous computation, for example by choosing the median as the threshold if the node sample size exceeds twenty.

Now, we construct a decision tree from the observed data. To this end, we prepare criteria for whether to continue branching. For example,

1. When the node sample size is less than n_{min},
2. When branching, no sample is contained in either of the group,
3. When the decrease before and after branching is less than an a priori determined threshold $(= \alpha)$.

The third item is due to the fact that the difference in (8.5) is the difference in the first terms subtracted by α.

When constructing a decision tree, we express each node as a list in the algorithm. For the inner nodes (excluding the terminal nodes but including the root), the attributes are the parent node, the left and right children nodes, the variable for branching, and the sample that is the threshold. For the terminal nodes, the attributes are the parent nodes and the response at the region. In the following procedure, the inner and terminal nodes contain as an attribute the set of the samples that reached the node.

The following procedure realizes stacking: the stack is empty at the beginning. When a node becomes an inner node, the left and right children are pushed to the stack (the stack height will increase by two), which means that they may branch in the future. Then, we take the node from the top of the stack (the stack height will decrease by one) and check whether it needs to branch. If it branches, the two children are placed on the top of the stack (the stack height will increase by two); otherwise, we check the node on the top of the current stack. The process continues until the stack is empty (Fig. 8.3). The stack itself is a list that stores the information about the parents of the nodes in the stack.

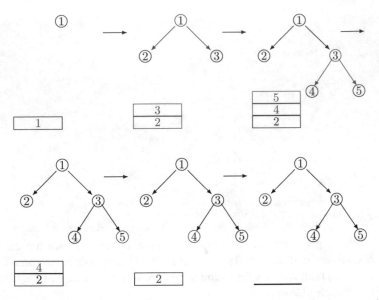

Fig. 8.3 We generate decision trees using a stack. Upper Left: 1 is pushed to the stack at the beginning. Upper Middle: 1 is removed from the stack, and 2, 3 are added to the stack. Upper Right: 3 is removed from the stack, and 4, 5 are added to the stack. Lower Left: 5 is removed from the stack. Lower Middle: 4 is removed from the stack. Lower Right: 2 is removed from the stack. In the decision trees, the red circles represent POP, and the blue lines express PUSH

```
class Stack:
    def __init__(self,parent,set,score):
        self.parent=parent
        self.set=set
        self.score=score
```

```
class Node:
    def __init__(self,parent,j,th,set):
        self.parent=parent
        self.j=j
        self.th=th
        self.set=set
```

```
def dt(x,y,alpha=0,n_min=1,rf=0):
    if rf==0:
        m=x.shape[1]
    # A single set of stack is constructed. Decision tree is initialized.
    stack=[Stack(0,list(range(x.shape[0])),f(y))]  # f is global
    node=[]
    k=-1
    #  Extracting the last element of the stack and updating the decision
        tree
    while len(stack)>0:
        popped=stack.pop()
        k=k+1
```

```
            i,j,left,right,score,left_score,right_score=branch(x,y,popped.set,rf
                )
            if popped.score-score<alpha or len(popped.set)<n_min or len(left)==0
                or len(right)==0:
                node.append(Node(popped.parent,-1,0,popped.set))
            else:
                node.append(Node(popped.parent,j,x[i,j],popped.set))
                stack.append(Stack(k,right,right_score))
                stack.append(Stack(k,left,left_score))
    # After these , set the value of node.left and node.right.
    for h in range(k,-1,-1):
        node[h].left=0; node[h].right=0;
    for h in range(k,0,-1):
        pa=node[h].parent
        if node[pa].right==0:
            node[pa].right=h
        else:
            node[pa].left=h
    # After these , calculate the value of node.center
    if f==sq_loss:
        g=np.mean
    else:
        g=mode_max
    for h in range(k+1):
        if node[h].j==-1:
            node[h].center=g(y[node[h].set])
        else:
            node[h].center=0
    return node
```

Each of the node [[1]] , node [[2]] , ... contains the left and right node IDs of its children as its attribute if they are inner nodes. The procedure adds the information giving the ID first to the parent, then to the left child, and finally to the right child. In each element of the output node, we have the variable for branching and its threshold for the inner nodes, the sample set in the node and the flag that the node is terminal for the terminal nodes, the parent node ID for the nodes except the root, and the left and right children node IDs for inner nodes. The attribute j expresses the variable for branching from 1 to p and is zero when the node is terminal. In the last stage, as an attribute of the list, the terminal nodes obtain the average of the responses in the region for regression and obtain the mode in the region for classification.

Example 64 (Boston Data Set) Data set for the Boston median housing prices (Response) and thirteen other covariates ($N = 506$). For $\alpha = 0$, $n.min = 50$, we construct a decision tree (Fig. 8.4). The procedure is implemented via the following code:

```
from sklearn.datasets import load_boston
```

```
boston=load_boston()
X=boston.data
y=boston.target
f=sq_loss
node=dt(X,y,n_min=50)
len(node)
```

Fig. 8.4 We construct a
decision tree with thirteen
covariates that explains the
median housing prices in
Boston. The thirteen variables
and their IDs are in the lower
left, and the IDs of the
branching variable and its
threshold is in the lower right

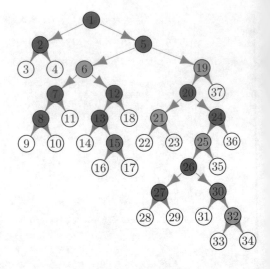

```
from igraph import *
```

```
r=len(node)
edge=[]
for h in range(1,r):
    edge.append([node[h].parent,h])
TAB=[];
for h in range(r):
    if not node[h].j==0:
        TAB.append([h,node[h].j,node[h].th])
TAB
```

```
def draw_graph(node):
    r=len(node)
    col=[]
    for h in range(r):
        col.append(node[h].j)
    colorlist=['#ffffff','#fff8ff','#fcf9ce','#d6fada','#d7ffff','#d9f2f8','
      #fac8be','#ffebff','#ffffe0','#fdf5e6','#fac8be','#f8ecd5','#ee82ee']
    color=[colorlist[col[i]] for i in range(r)]
    edge=[]
    for h in range(1,r):
        edge.append([node[h].parent,h])
        g=Graph(edges=edge,directed=True)
        layout=g.layout_reingold_tilford(root=[0])
    out=plot(g,vertex_size=15,layout=layout,bbox=(300,300),vertex_label=list
      (range(r)),vertex_color=color)
    return out
```

```
draw_graph(node)
```

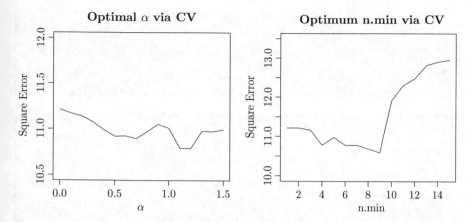

Fig. 8.5 We execute CV for the Boston data set (the first $N = 100$ data) and change the values of α to evaluate the CV values (Left). We observe that $\alpha = 0$ does not necessarily produce the best results. A value of approximately $1.0 \leq \alpha \leq 1.1$ is the best for CV. Additionally, we compute the CV evaluations while changing the *n.min* from 1 to 15 (Right). The best value is approximately $n.min = 9$

We wish to obtain the optimum α via CV for criterion (8.5). First, we construct the following value; then, we obtain the region R_j to which each $x \in \mathbb{R}^p$ belongs.

```
def value(u,node):
    r=0
    while node[r].j!=-1:
        if u[node[r].j]<node[r].th:
            r=node[r].left
        else:
            r=node[r].right
    return node[r].center
```

Example 65 For 10-fold CV and the Boston data set, we execute the procedure to obtain the optimum $0 \leq \alpha \leq 1.5$ (Fig. 8.5 Left). The execution is implemented via the following code. Because this process consumes considerable amounts of time, we execute only for the first $N = 100$ data.

```
boston=load_boston()
n=100
X=boston.data[range(n),:]
y=boston.target[range(n)]
f=sq_loss
alpha_seq=np.arange(0,1.5,0.1)
s=np.int(n/10)
out=[]
for alpha in alpha_seq:
    SS=0
    for h in range(10):
        test=list(range(h*s,h*s+s))
        train=list(set(range(n))-set(test))
        node=dt(X[train,:],y[train],alpha=alpha)
        for t in test:
```

```
                SS=SS+(y[t]-value(X[t,:],node))**2
    print(SS/n)
    out.append(SS/n)
plt.plot(alpha_seq,out)
plt.xlabel('alpha')
plt.ylabel('MSE')
plt.title("_optimal_alpha_by_CV_(N=100)")
```

```
boston=load_boston()
n=100
X=boston.data[range(n),:]
y=boston.target[range(n)]
n_min_seq=np.arange(1,13,1)
s=np.int(n/10)
out=[]
for n_min in n_min_seq:
    SS=0
    for h in range(10):
        test=list(range(h*s,h*s+s))
        train=list(set(range(n))-set(test))
        node=dt(X[train,:],y[train],n_min=n_min)
        for t in test:
            SS=SS+(y[t]-value(X[t,:],node))**2
    print(SS/n)
    out.append(SS/n)
plt.plot(n_min_seq,out)
plt.xlabel('n_min')
plt.ylabel('MSE')
plt.title("optimal_n_min_by_CV(N=100)")
```

We also perform similar executions to search for the best $1 \leq n.min \leq 15$ via CV (Figs. 8.5 Right).

8.2 Decision Tree for Classification

Regarding decision trees for classification, the same response is assigned to the covariates that belong to the same region. If we assign the class with the highest posterior probability to each region, the error probability will be minimized. Specifically, if we map from the p covariate values to one of $Y = 1, \ldots, K$, then if the simultaneous probability $f_{XY}(x, k)$ is given, the rule

$$x_i \in R_j \implies \hat{y}_i = \bar{y}_j$$

minimizes the average error probability

$$\sum_{k=1}^{K}\sum_{j=1}^{m}\int_{R_j} I(\bar{y}_j \neq k)f_{XY}(x,k)dx \,,$$

where $\bar{y}_j$ is the k that maximizes $\bar{y}_j := \dfrac{\int_{R_j} f(x,k)dx}{\sum_{h=1}^{K}\int_{R_h} f(x,h)dx}$ and $I(A)$ is an
indicator that takes a value of one if condition A holds and zero otherwise.

Let n_j be the sample size of region R_j, and let $\bar{y}_j$ be the mode of k such that $x_i \in R_j$ and $y_i = k$. Then, we define the rule such that

$$x_i \in R_j \Longrightarrow \hat{y}_i = \bar{y}_j$$

and choose $m \geq 1$ and $R_1, \ldots, R_m$ that minimizes

$$\sum_{j=1}^{m}\sum_{i:x_i \in R_j} I(y_i \neq \bar{y}_j) \,. \tag{8.6}$$

Furthermore, in classification, we are concerned with overfitting. For example, if the sample size is one for each region, the quantity of (8.6) is zero.

In addition, when we generate a decision tree for classification, we choose some criterion for branching according to a specific variable and threshold for the variable. If there are $n_{j,k}$ samples in region R_j such that $Y = k$, which can be expressed by $\sum_{j=1}^{m}(n_j - \max_k n_{j,k})$ with $\hat{p}_{j,k} := n_{j,k}/n_j$, it is sufficient to minimize the error probability

$$E_j := 1 - \max_k \hat{p}_{j,k}$$

for each R_j. However, if the tree is deep or if the number K of classes is large, choosing a variable based on minimizing the error probability is not always appropriate. In those cases, instead of the error probability E, either the Gini index

$$G_j := \sum_{k=1}^{K} \hat{p}_{j,k}(1 - \hat{p}_{j,k})$$

or the entropy

$$D_j := -\sum_{k=1}^{K} \hat{p}_{j,k} \log \hat{p}_{j,k}$$

can be used as a criterion.

If we use E_j, G_j, D_j as the criteria at each branch for classification, we can change the function sq_loss for regression as follows:

```python
def freq(y):
    y=list(y)
    return [y.count(i) for i in set(y)]
```

```python
# Mode
def mode(y):
    n=len(y)
    if n==0:
        return 0
    return max(freq(y))
```

```python
# error rate
def mis_match(y):
    return len(y)-mode(y)
```

```python
# Gini
def gini(y):
    n=len(y)
    if n==0:
        return 0
    fr=freq(y)
    return sum([fr[i]/n*(n-fr[i]) for i in range(len(fr))])
```

```python
# Entropy
def entropy(y):
    n=len(y)
    if n==0:
        return 0
    freq=[y.count(i) for i in set(y)]
    return np.sum([-freq[i]*np.log(freq[i]/n) for i in range(len(freq))])
```

Note that the three values are not normalized, i.e., multiplied by n_j in each region R_j, which is due to comparing the index values before and after branching because the former and latter consist of one and two regions.

Example 66 (Fisher's Iris) For the Fisher's Iris data set ($N = 150$, $p = 4$, Fig. 8.6), we compared the decision trees generated based on the error rate, Gini index, and entropy. The test is evaluated by the data used above, and overfitting is allowed. The error rate selects the variables that can distinguish the most frequent class from the other classes in the initial stage. On the other hand, the Gini index and entropy consider all classes and choose a tree that minimizes ambiguity (Figs. 8.7

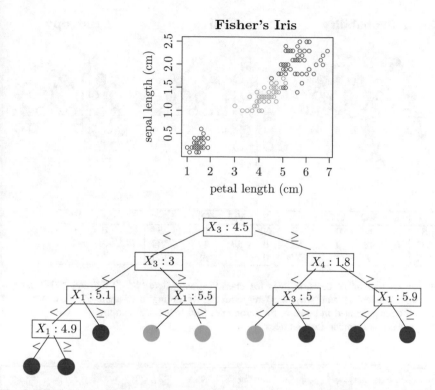

Fig. 8.6 Red, blue, and green circles express Iris setosa, Iris virginica, and Iris versicolor. Some samples of Iris virginica and Iris versicolor overlap (Upper), and if we generate a decision tree with $n.min = 20$ using all the 150 samples, eight inner and nine terminal nodes appear (Lower)

and 8.8). The execution of the process is implemented via the following code ($n.min = 4, \alpha = 0$):

```
def table_count(m,u,v):       # Again
    n=u.shape[0]
    count=np.zeros([m,m])
    for i in range(n):
        count[int(u[i]),int(v[i])]+=1
    return count
```

```
def mode_max(y):
    if len(y)==0:
        return -np.inf
    count=np.bincount(y)
    return np.argmax(count)
```

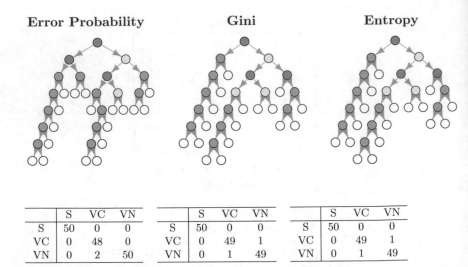

Error Probability				Gini				Entropy		
	S	VC	VN		S	VC	VN		S	VC	VN
S	50	0	0	S	50	0	0	S	50	0	0
VC	0	48	0	VC	0	49	1	VC	0	49	1
VN	0	2	50	VN	0	1	49	VN	0	1	49

Fig. 8.7 Generation of decision trees for classification, where "S," "VC," and "VN" express "Setosa," "Versicolor," and "Virginica," respectively. According to Example 66, we compare the decision trees generated in terms of the error rate, Gini index, and entropy. The Gini index and entropy generated similar decision trees

```
from sklearn.datasets import load_iris
```

```
iris=load_iris()
iris.target_names
f=mis_match
x=iris.data
y=iris.target
n=len(x)
node=dt(x,y,n_min=4)
m=len(node)
u=[]; v=[]
for h in range(m):
    if node[h].j==-1:
        w=y[node[h].set]
        u.extend([node[h].center]*len(w))
        v.extend(w)
table_count(3,np.array(u),np.array(v))
```

```
draw_graph(node)
```

Moreover, for classification, we can generate the optimum decision tree based on CV, as done for regression. For example, we may implement the following code:

```
iris=load_iris()
iris.target_names
f=mis_match
index=np.random.choice(n,n,replace=False) # Choose n from n candidates.
X=iris.data[index,:]
y=iris.target[index]
n_min_seq=np.arange(5,51,5)
s=15
for n_min in n_min_seq:
    SS=0
    for h in range(10):
        test=list(range(h*s,h*s+s))
        train=list(set(range(n))-set(test))
        node=dt(X[train,:],y[train],n_min=n_min)
        for t in test:
            SS=SS+np.sum(y[t]!=value(X[t,:],node))
    print(SS/n)
```

```
0.08666666666666667
0.08
0.07333333333333333
0.08
0.08
0.08
0.08
0.08
0.08
0.08
```

However, when the error rate (prediction performance) for new data is evaluated, the expected performance is not obtained (correct answer rate is approximately 90 %). To lower the classification error rate for future data, the K-nearest neighbor method (Chap. 2), logistic regression (Chap. 2), or support vector machine (Chap. 8) may be considered. However, as a generalization of decision tree regression and classification, we can generate multiple decision trees (random forest, boosting) and expect that these trees would have acceptable performance. These methods are described later in this chapter.

8.3 Bagging

Bagging applies the same idea as bootstrapping to the generation of decision trees: randomly select the same number of rows from a data frame (allow duplication) and use these data to generate a tree. This operation is repeated B times to obtain decision trees $\hat{f}_1, \ldots, \hat{f}_B$. Each tree takes the form of a function that performs regression or classification, and it takes a real value output for regression but a finite number of values prepared in advance for classification. When we obtain the outputs of the trees, $\hat{f}_1(x), \ldots, \hat{f}_B(x)$, for the new input $x \in \mathbb{R}^p$, the output is the arithmetic mean of the outputs and the value with the highest frequency for regression and for classification, respectively. Such processing is called bagging.

Consider two decision trees generated from the sample sets $(x_1, y_1), \ldots, (x_N, y_N)$ and $(x'_1, y'_1), \ldots, (x'_N, y'_N)$ whose distributions are shared. The sets often produce

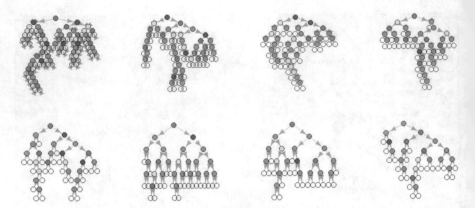

Fig. 8.8 We sample data from one data frame and a decision tree is generated for each data frame. Each decision tree has a similar function, but the variables selected may differ greatly. After generating a large number of such trees, we output the average and the most frequent class for regression and classification, respectively

completely different decision trees because the samples are easy to fit with different decision trees.

Therefore, when processing new data using the generated decision tree for either regression and classification, the result may be unreliable because the decision tree is unstable. Therefore, one approach is to generate many data frames, generate decision trees corresponding to the data frames, and obtain a solution under the consensus system of a plurality of decision trees.

Example 67 The decision trees in Fig. 8.8 are trees that actually sample data frames. The procedure uses the following code:

```
n=200
p=5
X=np.random.randn(n,p)
beta=randn(p)
Y=np.array(np.abs(np.dot(X,beta)+randn(n)),dtype=np.int64)
f=mis_match
node_seq=[]
for h in range(8):
    index=np.random.choice(n,n,replace=True) # Choose n from n candidates.
    x=X[index,:]
    y=Y[index]
    node_seq.append(dt(x,y,n_min=6))
```

```
draw_graph(node_seq[0])
```

```
draw_graph(node_seq[1])
```

8.4   Random Forest

Although bagging suppresses variation in the generated decision trees, the correlation among the generated decision trees is strong, and the original purpose is not sufficiently achieved. Therefore, an improved version called a random forest has been developed. The difference between bagging and random forest is that in random forest, the candidate variables used for branching are a subset of m variables instead of all p variables. The m variables are randomly selected for each branching, and the optimal variable is selected from among this subset. The theoretical consideration is beyond the scope of this book, but $m = \sqrt{p}$ is used in this section. The procedure that has been built thus far can be used to implement random forest by simply changing (generalizing) a part of the function branch. In the default situation of $m = p$, the procedure behaves the same as bagging, and the previous procedure also works.

```
def branch(x,y,S,rf=0):                                      ##
    if rf==0:                                                 ##
        T=np.arange(x.shape[1])                              ##
    else:                                                    ##
        T=np.random.choice(x.shape[1],rf,replace=False)      ##
    if x.shape[0]==0:
        return [0,0,0,0,0,0,0]
    best_score=np.inf
    for j in T:                                               ##
        for i in S:
            left=[]; right=[]
            for k in S:
                if x[k,j]<x[i,j]:
                    left.append(k)
                else:
                    right.append(k)
            left_score=f(y[left]); right_score=f(y[right])
            score=left_score+right_score
            if score<best_score:
                best_score=score
                i_1=i; j_1=j
                left_1=left; right_1=right
                left_score_1=left_score; right_score_1=right_score
    return [i_1,j_1,left_1,right_1,best_score,left_score_1,right_score_1]
```

For Fisher's Iris data set, the prediction was not correct when the decision tree was generated only once. In the case of random forest, however, we choose from $m \le p$ variables each time with the above function branch. Compared to the case of bagging, this approach produces a set of trees with large variations, which greatly improves the prediction performance.

Example 68 The trees are trained with 100 Iris training data points, and the performance is evaluated with 50 test data points. During the course of the experiment, the roles of the training and test data are not changed. The tree $b = 1, \ldots, B$ is generated each time, and the result of the i-th test data classification is stored in z[b, i].

For this two-dimensional array, we store the result of majority voting on b trees and the number of correct answers in $zz[b,i]$ and in $zzz[b]$, respectively, and define the function h that outputs the B-dimensional array.

```
def rf(z):
    z=np.array(z,dtype=np.int64)
    zz=[]
    for b in range(B):
        u=sum([mode_max(z[range(b+1),i])==y[i+100] for i in range(50)])
        zz.append(u)
    return zz
```

We execute the following program:

```
iris=load_iris()
iris.target_names
f=mis_match
n=iris.data.shape[0]
order=np.random.choice(n,n,replace=False) # Choose n from n candidates.
X=iris.data[order,:]
y=iris.target[order]
train=list(range(100))
test=list(range(100,150))
B=100
plt.ylim([35,55])
m_seq=[1,2,3,4]
c_seq=["r","b","g","y"]
label_seq=['m=1','m=2','m=3','m=4']
plt.xlabel('number_of_repeats')
plt.ylabel('the_number_of_correct_answers')
plt.title('random_forest')

for m in m_seq:
    z=np.zeros((B,50))
    for b in range(B):
        index=np.random.choice(train,100,replace=True)
        node=dt(X[index,:],y[index],n_min=2,rf=m)
        for i in test:
            z[b,i-100]=value(X[i,],node)
        plt.plot(list(range(B)),np.array(rf(z))-0.2*(m-2),label=label_seq[m-1],
        linewidth=0.8,c=c_seq[m-1])
plt.legend(loc='lower_right')
plt.axhline(y=50,c="b",linewidth=0.5,linestyle="dashed")
```

The results are shown in Fig. 8.9. It appears that $m = 4$ (bagging), where all variables are available, would be advantageous, but $m = 3$ had a lower error rate, and the performances of $m = 4$ and $m = 2$ were similar. In the case of bagging, only similar trees that make similar decisions are generated. Random forests have a certain probability of branching without using dominant variables, so trees that are different from bagging often occur and multifaceted decisions can be made, which is the advantage of a random forest.

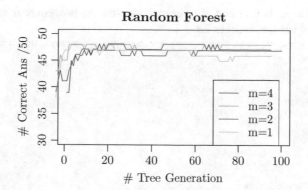

Fig. 8.9 Random forest applied to the Iris data set. When making decisions with a small number of decision trees, the error rate is large, even at $m = 4$ (bagging). The error rate improves as the number of trees generated increases, with the highest correct answer rate for $m = 3$, followed by $m = 2$. The lines are offset by 0.1 each to prevent the lines from overlapping

8.5 Boosting

The concept of boosting is broad, but we limit it to the case of using a decision tree. We select an appropriate $\lambda > 0$, limit the number of branches d (the number of end points is $d + 1$), and set $r = y$ (response) initially. Then, we generate trees and update the residual r sequentially as follows. We generate the tree $\hat{f}_1$ so that the difference between $[\hat{f}_1(x_1), \ldots, \hat{f}_1(x_N)]^T$ and r is the smallest, and we update $r = [r_1, \cdots, r_N]^T$ by

$$r_1 = r_1 - \lambda \hat{f}_1(x_1), \ldots, r_N = r_N - \lambda \hat{f}_1(x_N).$$

This process is repeated until $[\hat{f}_B(x_1), \ldots, \hat{f}_B(x_N)]^T$ is close to r. Finally, we generate the tree $\hat{f}_B$,

$$r_1 = r_1 - \lambda \hat{f}_B(x_1), \ldots, r_N = r_N - \lambda \hat{f}_B(x_N).$$

Here, the values of $r \in \mathbb{R}^N$ change as the tree generation progresses. Then, using the obtained $\hat{f}_1, \ldots, \hat{f}_B$, the final function

$$\hat{f}(\cdot) = \lambda \sum_{b=1}^{B} \hat{f}_b(\cdot)$$

is obtained.

First, we develop a procedure that generates appropriate trees given the number of inner points d (end points $d + 1$). The process b_dt is almost the same as dt: the only difference is that the number of interior points d (or the number of vertices $2d+1$) is predetermined. Therefore, branching starts from the vertex with the largest

difference in error before and after branching, and the procedure is stopped when
the number of vertices reaches $2d + 1$.

```
def b_dt(x,y,d):
    n=x.shape[0]
    node=[]
    first=Node(0,-1,0,np.arange(n))
    first.score=f(y[first.set])
    node.append(first)
    while len(node)<=2*d-1:
        r=len(node)
        gain_max=-np.inf
        for h in range(r):
            if node[h].j==-1:
                i,j,left,right,score,left_score,right_score=branch(x,y,node[
                    h].set)
                gain=node[h].score-score
                if gain>gain_max:
                    gain_max=gain
                    h_max=h
                    i_0=i; j_0=j
                    left_0=left; right_0=right
                    left_score_0=left_score; right_score_0=right_score
        node[h_max].th=x[i_0,j_0]; node[h_max].j=j_0
        next=Node(h_max,-1,0,left_0)
        next.score=f(y[next.set]); node.append(next)
        next=Node(h_max,-1,0,right_0)
        next.score=f(y[next.set]); node.append(next)
    r=2*d+1
    for h in range(r):
        node[h].left=0; node[h].right=0
    for h in range(r-1,1,-1):
        pa=node[h].parent
        if node[pa].right==0:
            node[pa].right=h
        else:
            node[pa].left=h
        if node[h].right==0 and node[h].left==0:
            node[h].j=-1
    if f==sq_loss:
        g=np.mean
    else:
        g=mode_max
    for h in range(r):
        if node[h].j==-1:
            node[h].center=g(node[h].set)
# After these , set the value of node.left and node.right.
    for h in range(r-1,-1,-1):
        node[h].left=0; node[h].right=0;
    for h in range(r-1,0,-1):
        pa=node[h].parent
        if node[pa].right==0:
            node[pa].right=h
        else:
            node[pa].left=h
# # After these , calculate the value of node.center
    if f==sq_loss:
        g=np.mean
    else:
        g=mode_max
    for h in range(r):
        if node[h].j==-1:
            node[h].center=g(y[node[h].set])
        else:
            node[h].center=0
    return node
```

For the choice of parameters d, B, λ, either $d = 1$ or $d = 2$ appears to be ideal.

Example 69 For the parameters B and λ, in general, it is necessary to decide the optimum for cross-validation, but since the implementation is in the Python language and runs slowly, in the following, we execute only the case $B = 200$ and $\lambda = 0.1$.

```
boston=load_boston()
B=200
lam=0.1
X=boston.data
y=boston.target
f=sq_loss
train=list(range(200))
test=list(range(200,300))
# Generate B boosting trees.
# It takes about 5 minutes for each d, for a total of about 15 minutes
trees_set=[]
for d in range(1,4):
    trees=[]
    r=y[train]
    for b in range(B):
        trees.append(b_dt(X[train,:],r,d))
        for i in train:
            r[i]=r[i]-lam*value(X[i,:],trees[b])
        print(b)
    trees_set.append(trees)
```

```
# Evaluation with test data
out_set=[]
for d in range(1,4):
    trees=trees_set[d-1]
    z=np.zeros((B,600))
    for i in test:
        z[0,i]=lam*value(X[i,],trees[0])
        for b in range(1,B):
            for i in test:
                z[b,i]=z[b-1,i]+lam*value(X[i,:],trees[b])
    out=[]
    for b in range(B):
        out.append(sum((y[test]-z[b,test])**2)/len(test))
    out_set.append(out)
```

```
# Displayed in graphs
plt.ylim([0,40])
c_seq=["r","b","g"]
label_seq=['d=1','d=2','d=3']
plt.xlabel('The_number_of_trees_generated')
plt.ylabel('MSE_with_test_data')
plt.title('This_book's program (lambda=0.1)')
for_d_in_range(1,4):
     out=out_set[d-1]
     u=range(20,100)
     v=out[20:100];
     plt.plot(u,v,label=label_seq[d-1],linewidth=0.8,c=c_seq[d-1])
plt.legend(loc='upper right')
```

We show how the square error (test data) changes in Fig. 8.10 for each $d = 1, 2, 3$ and $b = 1, \ldots, B$.

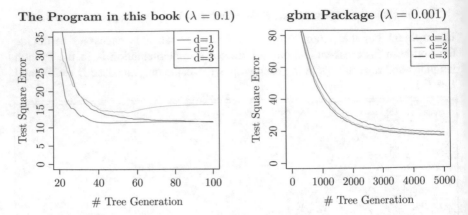

The Program in this book ($\lambda = 0.1$)   gbm Package ($\lambda = 0.001$)

Fig. 8.10 The executions of the program in this book with $\lambda = 0.1$ (Left) and gbm package with $\lambda = 0.001$ (Right). The smaller the λ is, the better the performance, but many trees should be generated

Gradient boosting essentially performs the above process. In actual processing, the `lightgbm` package is often used in Python language. Since this approach generates thousands of trees, it is devised for high speed processing.

Example 70 Setting $\lambda = 0.01$, $B = 5000$, $d = 1, 2, 3$, we execute the gradient boosting package `lightgbm` (Fig. 8.10 Right). Compared to the case of $\lambda = 0.1$, the results do not converge unless B is large. However, more accurate forecasts can be obtained. Additionally, the package is sufficiently fast: the procedure was completed in practical time even for $\lambda = 0.001$ (default) and $B = 5000$.

```
import lightgbm as lgb
```

```
boston=load_boston()
X=boston.data
y=boston.target
train=list(range(200))
test=list(range(200,300))
B=200
lgb_train=lgb.Dataset(X[train,:],y[train])
lgb_eval=lgb.Dataset(X[test,:],y[test],reference=lgb_train)
B=5000
nn_seq=list(range(1,10,1))+list(range(10,91,10))+list(range(100,B,50))
out_set=[]
for d in range(1,4):
    lgbm_params={
        'objective': 'regression',
        'metric': 'rmse',
        'num_leaves': d+1,
        'learning_rate': 0.001
    }
    out=[]
    for nn in nn_seq:
        model=lgb.train(lgbm_params,lgb_train,valid_sets=lgb_eval,
            verbose_eval=False,num_boost_round=nn)
```

```
        z=model.predict(X[test,:],num_iteration=model.best_iteration)
        out.append(sum((z-y[test])**2)/100)
    out_set.append(out)
```

```
# Displayed in graphs
plt.ylim([0,80])
c_seq=["r","b","g"]
label_seq=['d=1','d=2','d=3']
plt.xlabel('The_number_of_trees_generated')
plt.ylabel('MSE_with_test_data')
plt.title('light_(lambda=0.001)')
for d in range(1,4):
    out=out_set[d-1]
    u=range(20,100)
    plt.plot(nn_seq,out_set[d-1],label=label_seq[d-1],linewidth=0.8,c=c_seq[
        d-1])
plt.legend(loc='upper_right')
```

In this chapter, processing is implemented via the source program to understand the inner procedure of random forest and boosting, but in actual data analysis, such a package may be used.

Exercises 69–74

69. Write the following functions in the Python language, where each input y is a vector:

(a) sq_loss that given input vector y, outputs the square sum of the differences between each element and the arithmetic average.

(b) mis_match that given input vector y, outputs the number of mismatches between each element and the mode.

70. We used the function branch below to construct a tree. Given matrix x, vector y, loss function f, and the set of row indices S, the procedure outputs the division of S that minimizes the sum of the losses for the two new sets of indices. Fill in the blanks and execute the program.

```
def sq_loss(y):
    if len(y)==0:
        return 0
    else:
        y_bar=np.mean(y)
        return np.linalg.norm(y-y_bar)**2
```

```
def branch(x,y,S,rf=0):
    if rf==0:
        m=x.shape[1]
    if x.shape[0]==0:
        return [0,0,0,0,0,0,0]
    best_score=np.inf
```

```
    for j in range(x.shape[1]):
        for i in S:
            left=[]; right=[]
            for k in S:
                if x[k,j]<x[i,j]:
                    left.append(k)
                else:
                    # blank(1) #
            left_score=f(y[left]); right_score=f(y[right])
            score=# blank(2) #
            if score<best_score:
                best_score=score
                i_1=i; j_1=j
                left_1=left; right_1=right
                left_score_1=left_score; right_score_1=right_score
    return [i_1,j_1,left_1,right_1,best_score,left_score_1,right_score_1]
```

```
f=sq_loss
n=100; p=5
x=randn(n,p)
y=randn(n)
S=np.random.choice(n,10,replace=False)
branch(x,y,S)
```

71. The following procedure constructs a decision tree using the function branch and a loss function. Execute the procedure for Fisher's Iris data set and $n.min = 5$, $\alpha = 0$, and draw the graph.

```
class Stack:
    def __init__(self,parent,set,score):
        self.parent=parent
        self.set=set
        self.score=score
```

```
class Node:
    def __init__(self,parent,j,th,set):
        self.parent=parent
        self.j=j
        self.th=th
        self.set=set
```

```
def dt(x,y,alpha=0,n_min=1,rf=0):
    if rf==0:
        m=x.shape[1]
    # A single set of stack is constructed.Decision tree is initialized
    stack=[Stack(0,list(range(x.shape[0])),f(y))]   # f is global
    node=[]
    k=-1
    # Extracting the last element of the stack and updating the decision
      tree
    while len(stack)>0:
        popped=stack.pop()
        k=k+1
        i,j,left,right,score,left_score,right_score=branch(x,y,popped.set,
            rf)
        if popped.score-score<alpha or len(popped.set)<n_min or len(left
            )==0 or len(right)==0:
```

```
                    node.append(Node(popped.parent,-1,0,popped.set))
            else:
                    node.append(Node(popped.parent,j,x[i,j],popped.set))
                    stack.append(Stack(k,right,right_score))
                    stack.append(Stack(k,left,left_score))
    # After these , set the value of node.left and node.right.
    for h in range(k,-1,-1):
        node[h].left=0; node[h].right=0;
    for h in range(k,0,-1):
        pa=node[h].parent
        if node[pa].right==0:
            node[pa].right=h
        else:
            node[pa].left=h
    # After these , calculate the value of node.center
    if f==sq_loss:
        g=np.mean
    else:
        g=mode_max
    for h in range(k+1):
        if node[h].j==-1:
            node[h].center=g(y[node[h].set])
        else:
            node[h].center=0
    return node
```

The decision tree is obtained using below function if we get node.

```
from igraph import *
```

```
def draw_graph(node):
    r=len(node)
    col=[]
    for h in range(r):
        col.append(node[h].j)
    colorlist=['#ffffff','#fff8ff','#fcf9ce','#d6fada','#d7ffff','#d9f2f8',
        '#fac8be','#ffebff','#ffffe0','#fdf5e6','#fac8be','#f8ecd5','#ee82ee
        ']
    color=[colorlist[col[i]] for i in range(r)]
    edge=[]
    for h in range(1,r):
        edge.append([node[h].parent,h])
        g=Graph(edges=edge,directed=True)
        layout=g.layout_reingold_tilford(root=[0])
    out=plot(g,vertex_size=15,layout=layout,bbox=(300,300),vertex_label=
        list(range(r)),vertex_color=color)
    return out
```

```
draw_graph(node)
```

72. For the Boston data set, we consider finding the optimum $0 \le \alpha \le 1.5$ via 10-fold CV. Fill in either train or test in each blank to execute the procedure.

```
def value(u,node):
    r=0
    while node[r].j!=-1:
        if u[node[r].j]<node[r].th:
            r=node[r].left
```

```
            else:
                r=node[r].right
        return node[r].center
```

```
boston=load_boston()
n=100
X=boston.data[range(n),:]
y=boston.target[range(n)]
f=sq_loss
alpha_seq=np.arange(0,1.5,0.1)
s=np.int(n/10)
out=[]
for alpha in alpha_seq:
    SS=0
    for h in range(10):
        test=list(range(h*s,h*s+s))
        train=list(set(range(n))-set(test))
        node=dt(X[train,:],y[train],alpha=alpha)
        for t in test:
            SS=SS+(y[t]-value(X[t,:],node))**2
    print(SS/n)
    out.append(SS/n)
plt.plot(alpha_seq,out)
plt.xlabel('alpha')
plt.ylabel('MSE')
plt.title("_optimal_alpha_by_CV_(N=100)")
```

```
boston=load_boston()
n=100
X=boston.data[range(n),:]
y=boston.target[range(n)]
n_min_seq=np.arange(1,13,1)
s=np.int(n/10)
out=[]
for n_min in n_min_seq:
    SS=0
    for h in range(10):
        # blank #=list(range(h*s,h*s+s))
        # blank #=list(set(range(n))-set(# Blank #))
        node=dt(X[# blank #,:],y[# blank #],n_min=n_min)
        for t in # blank #:
            SS=SS+(y[t]-value(X[t,:],node))**2
    print(SS/n)
    out.append(SS/n)
plt.plot(n_min_seq,out)
plt.xlabel('n_min')
plt.ylabel('MSE')
plt.title("_optimal_n_min_by_CV_(N=100)")
```

73. We wish to modify `branch` and to construct a random forest procedure. Fill in the blanks, and execute the procedure.

```
def branch(x,y,S,rf=0):                                          ##
    if rf==0:                                                    ##
        T=# blank(1) #                                            ##
    else:                                                        ##
        T=# blank(2) #
    if x.shape[0]==0:
        return [0,0,0,0,0,0,0]
    best_score=np.inf
```

```
        for j in T:                                              ##
            for i in S:
                left=[]; right=[]
                for k in S:
                    if x[k,j]<x[i,j]:
                        left.append(k)
                    else:
                        right.append(k)
                left_score=f(y[left]); right_score=f(y[right])
                score=left_score+right_score
                if score<best_score:
                    best_score=score
                    i_1=i; j_1=j
                    left_1=left; right_1=right
                    left_score_1=left_score; right_score_1=right_score
        return [i_1,j_1,left_1,right_1,best_score,left_score_1,right_score_1]
```

```
def rf(z):
    z=np.array(z,dtype=np.int64)
    zz=[]
    for b in range(B):
        u=sum([mode_max(z[range(b+1),i])==y[i+100] for i in range(50)])
        zz.append(u)
    return zz
```

```
iris=load_iris()
iris.target_names
f=mis_match
n=iris.data.shape[0]
order=np.random.choice(n,n,replace=False) # Choose n from n candidates.
X=iris.data[order,:]
y=iris.target[order]
train=list(range(100))
test=list(range(100,150))
B=100
plt.ylim([35,55])
m_seq=[1,2,3,4]
c_seq=["r","b","g","y"]
label_seq=['m=1','m=2','m=3','m=4']
plt.xlabel('the_number_of_repeats')
plt.ylabel('the_number_of_correct_answers')
plt.title('random_forest')
for m in m_seq:
    z=np.zeros((B,50))
    for b in range(B):
        index=np.random.choice(train,100,replace=True)
        node=dt(X[index,:],y[index],n_min=2,rf=m)
        for i in test:
            z[b,i-100]=value(X[i,],node)
    plt.plot(list(range(B)),np.array(rf(z))-0.2*(m-2),label=label_seq[m-1],
        linewidth=0.8,c=c_seq[m-1])
plt.legend(loc='lower_right')
plt.axhline(y=50,c="b",linewidth=0.5,linestyle="dashed")
```

74. We execute boosting using the `lightgbm` package for the Boston data set.
    Look up the `lightgbm` package, fill in the blanks, and draw the graph.

```
import lightgbm as lgb
```

```
boston=load_boston()
X=boston.data
y=boston.target
train=list(range(200))
test=list(range(200,300))
B=200
lgb_train=lgb.Dataset(X[train,:],y[train])
lgb_eval=lgb.Dataset(X[test,:],y[test],reference=lgb_train)
B=5000
nn_seq=list(range(1,10,1))+list(range(10,91,10))+list(range(100,B,50))
out_set=[]
for d in range(1,4):
    lgbm_params={
        'objective': 'regression',
        'metric': 'rmse',
        'num_leaves': # blank(1) #,
        'learning_rate': 0.001
    }
    out=[]
    for nn in nn_seq:
        model=lgb.train(lgbm_params,lgb_train,valid_sets=lgb_eval,
            verbose_eval=False,num_boost_round=# blank(2) #)
        z=model.predict(X[test,:],num_iteration=model.best_iteration)
        out.append(sum((z-y[test])**2)/100)
    out_set.append(out)
```

```
# Displayed in graphs
plt.ylim([0,80])
c_seq=["r","b","g"]
label_seq=['d=1','d=2','d=3']
plt.xlabel('The_number_of_trees_generated')
plt.ylabel('MSE_with_test_data')
plt.title('lightgbm_package_(lambda=0.001)')
for d in range(1,4):
    out=out_set[d-1]
    u=range(20,100)
    v=out[20:100];
    plt.plot(nn_seq,out_set[d-1],label=label_seq[d-1],linewidth=0.8,c=c_seq
        [d-1])
plt.legend(loc='upper_right')
```

Chapter 9
Support Vector Machine

Abstract Support vector machine is a method for classification and regression that draws an optimal boundary in the space of covariates (p dimension) when the samples $(x_1, y_1), \ldots, (x_N, y_N)$ are given. This is a method to maximize the minimum value over $i = 1, \ldots, N$ of the distance between x_i and the boundary. This notion is generalized even if the samples are not separated by a surface by softening the notion of a margin. Additionally, by using a general kernel that is not the inner product, even if the boundary is not a surface, we can mathematically formulate the problem and obtain the optimum solution. In this chapter, we consider only the two-class case and focus on the core part. Although omitted here, the theory of support vector machines also applies to regression and classification with more than two classes.

9.1 Optimum Boarder

In the following, we consider a classification rule with two classes from N samples, where the responses take the values of $y_1, \ldots, y_N = \pm 1$. To consider the locations of covariates geometrically, we first consider the distance between a point and a line.

Proposition 23 *The distance between $(x, y) \in \mathbb{R}^2$ and the line $l : aX + bY + c = 0$, $a, b \in \mathbb{R}$ is given by*

$$\frac{|ax + by + c|}{\sqrt{a^2 + b^2}}.$$

For the proof, see the Appendix at the end of this chapter.

The formula assumes $p = 2$-dimensional Euclidean space. For the general p-dimensional case, the distance between $x = [x_1, \ldots, x_p]$ (row vector) and the surface $\beta_0 + \beta_1 X_1 + \cdots + \beta_p X_p = 0$ is

$$d(x) := \frac{|\beta_0 + x_1\beta_1 + \cdots + x_p\beta_p|}{\sqrt{\beta_1^2 + \cdots + \beta_p^2}}.$$

© The Author(s), under exclusive license to Springer Nature Singapore Pte Ltd. 2021
J. Suzuki, *Statistical Learning with Math and Python*,
https://doi.org/10.1007/978-981-15-7877-9_9

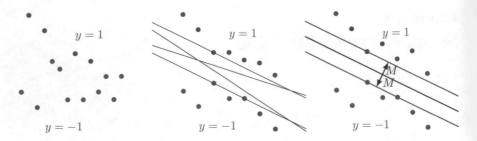

Fig. 9.1 Left: the samples are not separable by any surface. Middle: the samples are separable by a surface, but such surfaces are not unique. Right: the two parallel surfaces in red and blue do not contain any sample between them and they maximize the distance (twice M) between them. The middle (center) surface is the border between the samples such that $y_i = 1$ and $y_i = -1$. Only the samples on the red and blue surfaces determine the border; the others do not play such a role

In particular, if we divide each term of $\beta_0 \in \mathbb{R}$, $\beta = [\beta_1, \ldots, \beta_p]^T \in \mathbb{R}^p$ by the same constant such that $\|\beta\|_2 = 1$, we can write the result as

$$d(x) = |\beta_0 + x_1\beta_1 + \cdots + x_p\beta_p| .$$

In the following, we say that the samples are *separable* (by a surface) if there exist β_0 and β (Fig. 9.1) such that

$$y_1(\beta_0 + x_1\beta), \ldots, y_N(\beta_0 + x_N\beta) \geq 0 .$$

If the samples are separable, such a surface is not unique. In order to specify it, we define the following rule. Even if the samples are separable, we prepare two parallel surfaces that do not contain any sample between them. Then, we maximize the distance between the surfaces and regard the surface in the middle of the two surfaces as the border that divides the samples.

To this end, for the N samples $(x_1, y_1), \ldots, (x_N, y_N)$, it is sufficient to maximize the minimum distance $M := \min_i d(x_i)$ between each x_i and the surface $\beta_0 + X_1\beta_1 + \cdots + X_p\beta_p = 0$. Without loss of generality, if we assume that the coefficients β_0 and β satisfy $\|\beta\|_2 = 1$, because $d(x_i) = y_i(\beta_0 + x_i\beta)$, the problem reduces to finding the β_0, β that maximize the *margin*

$$M := \min_{i=1,\ldots,N} y_i(\beta_0 + x_i\beta)$$

(Fig. 9.1). In this case, the subset (support vector) of $\{1, \ldots, N\}$ such that $M = y_i(\beta_0 + x_i\beta)$ determines the coefficients β_0 and β and the margin M.

Even if the samples are separable, another set of N samples that follow the same distribution may not be separable. If we formulate the problem of obtaining the border (surface) from the samples in the general setting, we should assume that the samples are separable. Rather, we define a formulation for any samples that are not necessarily separable. Now, we generalize the result as follows. Given $\gamma \geq 0$, we maximize M w.r.t. $(\beta_0, \beta) \in \mathbb{R} \times \mathbb{R}^p$ and $\epsilon_i \geq 0, i = 1, \ldots, N$, under the constraints

$$\sum_{i=1}^{N} \epsilon_i \leq \gamma \tag{9.1}$$

and

$$y_i(\beta_0 + x_i\beta) \geq M(1 - \epsilon_i). \tag{9.2}$$

$i = 1, \ldots, N$.

For separable samples, we solve the problem under the setting $\epsilon_1 = \cdots = \epsilon_N = 0$, i.e., solve it for $\gamma = 0$ (Fig. 9.2, left). However, we may formulate the problem $\gamma > 0$ even if the samples are separable. In that case, the value of M increases because the constraint is relaxed.

The i's such that $\epsilon_i > 0$, as well as those such that $M = y_i(\beta_0 + x_i\beta)$ are the support vectors. In other words, all the i values that satisfy (9.2) with equality, are the support vectors for the margin M. Compared to the case of $\gamma = 0$, because the support vectors increase, more samples support the decision of β_0 and β for the optimal border (Fig. 9.2, middle). For this case, the estimation is less sensitive to the variation of samples, which means that we need to adjust the value of γ appropriately, for example, via cross-validation.

On the other hand, the values of ϵ_i are $\epsilon = 0, 0 < \epsilon < 1, \epsilon = 1$, and $\epsilon > 1$ for $y_i(\beta_0 + x_i\beta) \geq M, 0 < y_i(\beta_0 + x_i\beta) < M, y_i(\beta_0 + x_i\beta) = 0$, and $y_i(\beta_0 + x_i\beta) < 0$, respectively.

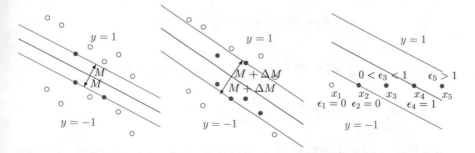

Fig. 9.2 The samples filled in either red or blue are the support vectors. Left: there exists a solution of $\gamma = 0$ for separable samples. In this example, the border is determined by only the three samples (support vectors). However, if we set $\gamma > 0$ (middle), we allow samples to be between the red and blue surfaces because the margin becomes large, but the solution is stable because the six support vectors support the red surface. When the value of γ changes, the support vectors change, so does the border surface. Right: for $\{x_1, x_2\}, x_3, x_4$, and x_5, we observe $\epsilon = 0, 0 < \epsilon < 1, \epsilon = 1$, and $\epsilon > 1$, respectively. If $\epsilon = 0$, some are support vectors, as for x_1, and others are not, as for x_2

The conditions $\epsilon_i = 1$ and $0 < \epsilon < 1$ are on the border and between the border and the margin in front. Some of the i's such that $\epsilon_i = 0$ are on the margin in front and others are not (support vectors) (Fig. 9.2, right).

For nonseparable samples, if the value of γ is too small, no solution exists. In fact, if $\gamma = 0$ and at least one i such that $\epsilon_i = 0$ does not satisfy (9.2), then no solution exists. On the other hand, if the number of such i values for which $\epsilon_i > 1$ exceeds γ, no such β_0 and β exist.

The support vector machine problem is formulated as finding $\beta_0, \beta, \epsilon_i, i = 1, \ldots, N$, that minimize

$$L_P := \frac{1}{2}\|\beta\|^2 + C\sum_{i=1}^{N}\epsilon_i - \sum_{i=1}^{N}\alpha_i\{y_i(\beta_0 + x_i\beta) - (1 - \epsilon_i)\} - \sum_{i=1}^{N}\mu_i\epsilon_i , \quad (9.3)$$

removing the constraint $\|\beta\|_2 = 1$ and replacing β_0/M, β/M by β_0, β in (9.2). Note that minimizing $\|\beta\|$ is equivalent to maximizing the M before β is normalized as β/M. In this setting, we regard $C > 0$ as a cost, the last two terms are constraints, and $\alpha_i, \mu_i \geq 0, i = 1, \ldots, N$, are the Lagrange coefficients.

9.2 Theory of Optimization

Before solving (9.1) and (9.2), we prepare the theory. In the following, in the problem of finding $\beta \in \mathbb{R}^p$ that minimizes $f_0(\beta)$ under $f_j(\beta) \leq 0, \ j = 1, \ldots, m$, we assume that such a solution exists and that such a β is β^*.

We define

$$L(\alpha, \beta) := f_0(\beta) + \sum_{j=1}^{m}\alpha_j f_j(\beta)$$

for $\alpha = (\alpha_1, \ldots, \alpha_m) \in \mathbb{R}^m$. Then, for an arbitrary[1] $\beta \in \mathbb{R}^p$, we have

$$\sup_{\alpha \geq 0} L(\alpha, \beta) = \begin{cases} f_0(\beta), & f_1(\beta) \leq 0, \ldots, f_m(\beta) \leq 0 \\ +\infty, & \text{Otherwise.} \end{cases} \quad (9.4)$$

In fact, for an arbitrarily fixed $\beta \in \mathbb{R}^p$, if j exists such that $f_j(\beta) > 0$ and we make α_j larger, then $L(\alpha, \beta)$ can be infinitely large. On the other hand, if

---

[1]We say that v is an upper bound of S if $u \leq v$ for any $u \in S$ in a set $S \subseteq \mathbb{R}$ and that the minimum of the upper bounds of S is the upper limit of S, which we write as sup A. For example, the maximum does not exist for $S = \{x \in \mathbb{R}|0 \leq x < 1\}$, but sup $S = 1$. Similarly, we define lower bounds and their maximum (the upper limit) of S, which we write as inf A.

$f_1(\beta), \ldots, f_m(\beta) \leq 0$, then $L(\alpha, \beta)$ has the largest value $f_0(\beta)$ for $\alpha \geq 0$ when $\alpha_1 = \cdots = \alpha_m = 0$.

Moreover, we have

$$f^* := \inf_{\beta} \sup_{\alpha \geq 0} L(\alpha, \beta) \geq \sup_{\alpha \geq 0} \inf_{\beta} L(\alpha, \beta) . \qquad (9.5)$$

In fact, for arbitrary $\alpha' \geq 0$ and $\beta' \in \mathbb{R}^p$, we have

$$\sup_{\alpha \geq 0} L(\alpha, \beta') \geq L(\alpha', \beta') \geq \inf_{\beta} L(\alpha', \beta) .$$

Since the inequality holds for arbitrary α' and β', it still holds even if we take the inf and sup w.r.t. the β' and α' on the left and right, respectively.

Example 71 Suppose that $p = 2$, $m = 1$, $f_0(\beta) := \beta_1 + \beta_2$, and $f_1(\beta) = \beta_1^2 + \beta_2^2 - 1$. Then, for

$$L(\alpha, \beta) := \beta_1 + \beta_2 + \alpha(\beta_1^2 + \beta_2^2 - 1) ,$$

we write (9.4) as

$$\sup_{\alpha \geq 0} L(\alpha, \beta) = \begin{cases} \beta_1 + \beta_2, & \beta_1^2 + \beta_2^2 \leq 1 \\ +\infty, & \text{Otherwise.} \end{cases}$$

Thus, it takes the minimum value $-\sqrt{2}$ when $\beta_1 = \beta_2 = -1/\sqrt{2}$, and the left-hand side of (9.5) is $\sqrt{2}$. If we partially differentiate $L(\alpha, \beta)$ w.r.t. β_1, β_2, we have $\beta_1 = \beta_2 = -1/(2\alpha)$ and

$$\inf_{\beta} L(\alpha, \beta) = -\frac{1}{2\alpha} - \frac{1}{2\alpha} + \alpha \left\{ \left(\frac{1}{2\alpha}\right)^2 + \left(\frac{1}{2\alpha}\right)^2 - 1 \right\} = -\frac{1}{2\alpha} - \alpha .$$

From the inequality between the arithmetic and geometric means, the value is maximized at $1/(2\alpha) = \alpha$. Therefore, from $\alpha = -1/\sqrt{2}$, the right-hand side of (9.5) is $-\sqrt{2}$ as well.

The problem of minimizing $f(\beta) := \sup_{\alpha \geq 0} L(\alpha, \beta)$ is a primary problem, while that of maximizing $g(\alpha) := \inf_{\beta} L(\alpha, \beta)$ under $\alpha \geq 0$ is a dual problem.

If we write the optimum values of the primary and dual problems as $f^* := \inf_{\beta} f(\beta)$ and $g^* := \sup_{\alpha \geq 0} g(\alpha)$, then we have

$$f^* \geq g^*. \qquad (9.6)$$

We consider only the case that the inequality is equal in this book. Many problems, including support vector machines, satisfy this assumption.

Assumption 1 $f^* = g^*$.

Suppose that $f_0, f_1, \ldots, f_m : \mathbb{R}^p \to \mathbb{R}$ are convex and differentiable at $\beta = \beta^*$. We consider the equivalent conditions in the following proposition below according to the KKT (Karush–Kuhn–Tucker) conditions.

Proposition 24 (KKT Condition) *Under* $f_1(\beta) \leq 0, \ldots, f_m(\beta) \leq 0$, *the solution* $\beta = \beta^* \in \mathbb{R}^p$ *minimizes* $f_0(\beta)$ *if and only if there exist* $\alpha_1, \ldots, \alpha_m \geq 0$ *s.t.*

$$f_1(\beta^*), \ldots, f_m(\beta^*) \leq 0 \tag{9.7}$$

$$\alpha_1 f_1(\beta^*) = \cdots = \alpha_m f_m(\beta^*) = 0 \tag{9.8}$$

$$\nabla f_0(\beta^*) + \sum_{i=1}^{m} \alpha_i \nabla f_i(\beta^*) = 0. \tag{9.9}$$

Proof Sufficiency: as proved for $p = 1$ in Chap. 6, in general, for any $p \geq 1$, if $f : \mathbb{R}^p \to \mathbb{R}$ is convex and differentiable at $x = x_0 \in \mathbb{R}^p$, for each $x \in \mathbb{R}^p$, we have

$$f(x) \geq f(x_0) + \nabla f(x_0)^T (x - x_0) . \tag{9.10}$$

Using the fact that, under the KKT conditions, for each solution $\beta \in \mathbb{R}^p$, we have

$$f_0(\beta^*) \leq f_0(\beta) - \nabla f_0(\beta^*)^T (\beta - \beta^*) = f_0(\beta) + \sum_{i=1}^{m} \alpha_i \nabla f_i(\beta^*)^T (\beta - \beta^*)$$

$$\leq f_0(\beta) + \sum_{i=1}^{m} \alpha_i \{ f_i(\beta) - f_i(\beta^*) \} = f_0(\beta) + \sum_{i=1}^{m} \alpha_i f_i(\beta) \leq f_0(\beta) ,$$

which means that β^* is optimum, and we used (9.10), (9.9), (9.10), (9.8), and $f_1(\beta) \leq 0, \ldots, f_m(\beta) \leq 0$ in the above derivation steps.   □

Necessity: if β^* is the solution of the primary problem, then a solution β exists such that $f_1(\beta), \ldots, f_m(\beta) \leq 0$, and we require (9.7). From the assumption $f^* = g^*$, there exists $\alpha^* \geq 0$ such that

$$f_0(\beta^*) = g(\alpha^*) = \inf_{\beta} \left\{ f_0(\beta) + \sum_{i=1}^{m} \alpha_i^* f_i(\beta) \right\} \leq f_0(\beta^*) + \sum_{i=1}^{m} \alpha_i^* f_i(\beta^*) \leq f_0(\beta^*) ,$$

where the first equality is due to Assumption 1 and the last inequality is due to (9.7) and $\alpha^* \geq 0$. Thus, the above inequality is an equality, and we have (9.8). Finally, β^* minimizes

$$f_0(\beta) + \sum_{i=1}^{m} \alpha_i^* f_i(\beta),$$

and we require (9.9).

Example 72 For Example 71, the KKT conditions (9.7), (9.8), and (9.9) are

$$\beta_1^2 + \beta_2^2 - 1 \leq 0 \tag{9.11}$$

$$\alpha(\beta_1^2 + \beta_2^2 - 1) = 0 \tag{9.12}$$

$$\begin{bmatrix} 1 \\ 1 \end{bmatrix} + 2\alpha \begin{bmatrix} \beta_1 \\ \beta_2 \end{bmatrix} = \begin{bmatrix} 0 \\ 0 \end{bmatrix}, \tag{9.13}$$

where $\alpha = 0$ satisfies (9.12) but does not satisfy (9.13). Thus, the equalities in (9.11) and (9.13) are the KKT conditions.

9.3 The Solution of Support Vector Machines

The following seven equations are the KKT conditions:

$$y_i(\beta_0 + x_i\beta) - (1 - \epsilon_i) \geq 0 \tag{9.14}$$

$$\epsilon_i \geq 0 \tag{9.15}$$

are from (9.7), and

$$\alpha_i[y_i(\beta_0 + x_i\beta) - (1 - \epsilon_i)] = 0 \tag{9.16}$$

$$\mu_i\epsilon_i = 0 \tag{9.17}$$

are from (9.8). Moreover, from (9.9), differentiating (9.3) w.r.t. β, β_0, and ϵ_i, we obtain

$$\beta = \sum_{i=1}^{N} \alpha_i y_i x_i^T \in \mathbb{R}^p \tag{9.18}$$

$$\sum_{i=1}^{N} \alpha_i y_i = 0 \tag{9.19}$$

$$C - \alpha_i - \mu_i = 0, \tag{9.20}$$

respectively.

The dual problem of L_P in (9.3) is constructed as follows. In order to optimize w.r.t. β_0, ϵ_i, if we differentiate by these values, we obtain (9.19) and (9.20); thus, L_p can be stated as

$$\sum_{i=1}^{N} \alpha_i + \frac{1}{2}\|\beta\|^2 - \sum_{i=1}^{N} x_i \beta \alpha_i y_i.$$

From (9.18), the second and third terms become

$$\frac{1}{2}\left(\sum_{i=1}^{N} \alpha_i y_i x_i^T\right)^T \left(\sum_{i=1}^{N} \alpha_i y_i x_i^T\right) - \sum_{i=1}^{N} x_i \left(\sum_{i=1}^{N} \alpha_i y_i x_i^T\right) \alpha_i y_i \ ,$$

so we construct the function with input as the Lagrange coefficients $\alpha_i, \mu_i \geq 0$, $i = 1, \ldots, N$,

$$L_D := \sum_{i=1}^{N} \alpha_i - \frac{1}{2} \sum_{i=1}^{N} \sum_{j=1}^{N} \alpha_i \alpha_j y_i y_j x_i x_j^T \ , \tag{9.21}$$

where α ranges over (9.19) and

$$0 \leq \alpha_i \leq C \ . \tag{9.22}$$

Note that although μ_i is not included in L_D, $\mu_i = C - \alpha_i \geq 0$ and $\alpha_i \geq 0$ are left as (9.22). We can compute β via (9.18) from the α obtained in this manner.

In solving the dual problem, we note that (9.22) can be divided into the three cases.

Proposition 25

$$\begin{cases} \alpha_i = 0 & \Longleftarrow y_i(\beta_0 + x_i\beta) > 1 \\ 0 < \alpha_i < C \Longrightarrow y_i(\beta_0 + x_i\beta) = 1 \\ \alpha_i = C & \Longleftarrow y_i(\beta_0 + x_i\beta) < 1. \end{cases} \tag{9.23}$$

For the proof, see the end of this chapter.

We show that at least one i satisfies $y_i(\beta_0 + x_i\beta) = 1$.

If there exists an i such that $0 < \alpha_i < C$, then from Proposition 25, the i satisfies $y_i(\beta_0 + x_i\beta) = 1$, and we obtain β_0 via $\beta_0 = y_i - x_i\beta$.

Suppose $\alpha_1 = \cdots = \alpha_N = 0$. From (8.18), we have $\beta = 0$. Moreover, from (8.17) and (8.20), we have $\mu_i = C$ and $\epsilon_i = 0$, $i = 1, \cdots, N$, which means $y_i(\beta_0 + x_i\beta) \geq 1$, from Proposition 25. Therefore, we require $y_i\beta_0 \geq 1$, i.e., $y_1 = \cdots = y_N$. Then, $\beta_0 = 1$ and $\beta_0 = -1$ satisfy $y_i(\beta_0 + x_i\beta) = 1$ when $y_i = 1$ and $y_i = -1$, respectively.

Suppose next that $\alpha_i = C$ for at least one i (we denote the set of such i by S) and that $\alpha_i = 0$ for $\{1, \cdots, N\} \backslash S$. From (8.16), we have $y_i(\beta_0 + x_i\beta) - (1 - \epsilon_i) = 0$. In the following, we show that $\epsilon_i = 0$ for at least one $i \in S$. To this end, we assume $\epsilon_i > 0$ for all $i \in S$. If we define $\epsilon_* := \min_{i \in S} \epsilon_i$, and replace ϵ_i and β_0 by $\epsilon_* - \epsilon_*$ and $\beta_0 + y_i \epsilon_*$, respectively, the latter satisfies the constraint (8.14) and has a smaller value of

$$f_0(\beta, \beta_0, \epsilon) = \frac{1}{2}\|\beta\|^2 + C \sum_{i=1}^{N} \epsilon_i \,,$$

which contradicts the underlying assumption that β, β_0, and ϵ were optimal. Thus, we have $\epsilon_i = 0$ for at least one $i \in S$. Then, we can compute $\beta_0 = y_i - x_i\beta$, and ϵ_j for $j \in S \backslash \{i\}$. For each $i \in S$, we set $\epsilon_i = 0$ and compute $\sum_{j=1}^{N} \epsilon_j$. We choose the $i(\epsilon_i = 0)$ that minimizes the quantity and set such β_0 as the final β_0. The β, β_0, and ϵ minimize $f_0(\beta, \beta_0, \epsilon)$.

We solve the dual problem (9.19), (9.21), and (9.22) using a quadratic programming solver. In the R language, a package called $\{\texttt{quadprog}\}$ is available for this purpose. We specify $D_{mat} \in \mathbb{R}^{N \times N}$, $A_{mat} \in \mathbb{R}^{m \times N}$, $d_{vec} \in \mathbb{R}^N$, $b_{vec} \in \mathbb{R}^m$ $(m \geq 1)$ such that

$$L_D = -\frac{1}{2}\alpha^T D_{mat}\alpha + d_{vec}^T \alpha$$

$$A_{mat}\alpha \geq b_{vec} \,,$$

for $\alpha \in \mathbb{R}^N$, where we assume that the first meq and $m - med$ are equality and inequality constraints in $A_{mat}\alpha \geq b_{vec}$ and specify the number $0 \leq meq \leq m$. In particular, in the formulation derived above, we take $m = 2N + 1$, $meq = 1$,

$$z = \begin{bmatrix} x_{1,1}y_1 & \cdots & x_{1,p}y_1 \\ \vdots & \vdots & \vdots \\ x_{N,1}y_N & \cdots & x_{N,p}y_N \end{bmatrix} \in \mathbb{R}^{N \times p}, \quad A_{mat} = \begin{bmatrix} y_1 & \cdots & y_N \\ -1 & \cdots & 0 \\ 0 & \ddots & 0 \\ 0 & \cdots & -1 \\ 1 & \cdots & 0 \\ 0 & \ddots & 0 \\ 0 & \cdots & 1 \end{bmatrix} \in \mathbb{R}^{(2N+1) \times N}$$

$D_{mat} = zz^T \in \mathbb{R}^{N \times N}$ (if the rank is below N, the matrix is singular), $b_{vec} = [0, -C, \ldots, -C, 0, \ldots, 0]^T \in \mathbb{R}^{2N+1}$, $d_{vec} = [1, \ldots, 1]^T \in \mathbb{R}^N$, and $\alpha = [\alpha_1, \ldots, \alpha_N] \in \mathbb{R}^N$. For example, we construct the following procedure:

```
import cvxopt
from cvxopt import matrix
```

```
a=randn(1); b=randn(1)
n=100
X=randn(n,2)
y=np.sign(a*X[:,0]+b*X[:,1]+0.1*randn(n))
y=y.reshape(-1,1)   # The shape needs to be clearly marked
```

```
def svm_1(X,y,C):
    eps=0.0001
    n=X.shape[0]
    P=np.zeros((n,n))
    for i in range(n):
        for j in range(n):
            P[i,j]=np.dot(X[i,:],X[j,:])*y[i]*y[j]
    # It must be specified using the matrix function in cvxopt.
    P=matrix(P+np.eye(n)*eps)
    A=matrix(-y.T.astype(np.float))
    b=matrix(np.array([0]).astype(np.float))
    h=matrix(np.array([C]*n+[0]*n).reshape(-1,1).astype(np.float))
    G=matrix(np.concatenate([[np.diag(np.ones(n)),np.diag(-np.ones(n))]]))
    q=matrix(np.array([-1]*n).astype(np.float))

    res=cvxopt.solvers.qp(P,q,A=A,b=b,G=G,h=h)      # execute solver
    alpha=np.array(res['x'])   # where x corresponds to alpha in the text
    beta=((alpha*y).T@X).reshape(2,1)
    index=np.arange(0,n,1)
    index (eps<alpha[:,0])&(alpha[:,0]<c-eps)
    beta_0=np.mean(y[index]-X[index,:]@beta)

    return {'beta':beta,'beta_0':beta_0}
```

Example 73 Using the function svm_1, we execute the following procedure and drew the samples and the border as in Fig. 9.3:

```
a=randn(1); b=randn(1)
n=100
X=randn(n,2)
y=np.sign(a*X[:,0]+b*X[:,1]+0.1*randn(n))
y=y.reshape(-1,1)   # The shape needs to be clearly marked
for i in range(n):
    if y[i]==1:
        plt.scatter(X[i,0],X[i,1],c="red")
    else :
        plt.scatter(X[i,0],X[i,1],c="blue")
res=svm_1(X,y,C=10)
```

Fig. 9.3 Generating samples, we draw the border of the support vector machine

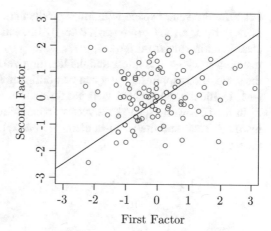

```
def f(x):
    return -res['beta_0']/res['beta'][1]-x*res['beta'][0]/res['beta'][1]
```

```
x_seq=np.arange(-3,3,0.5)
plt.plot(x_seq,f(x_seq))
res
```

```
        pcost         dcost        gap     pres    dres
 0: -1.6933e+02  -7.9084e+03   2e+04   8e-01   8e-15
 1: -1.4335e+01  -2.5477e+03   4e+03   1e-01   1e-14
 2:  3.4814e+01  -3.6817e+02   5e+02   1e-02   4e-14
 3: -2.0896e+01  -1.3363e+02   1e+02   3e-03   2e-14
 4: -4.4713e+01  -1.0348e+02   6e+01   1e-03   8e-15
 5: -5.8178e+01  -8.1212e+01   2e+01   4e-04   6e-15
 6: -6.4262e+01  -7.5415e+01   1e+01   1e-04   4e-15
 7: -6.7750e+01  -7.0997e+01   3e+00   2e-05   5e-15
 8: -6.9204e+01  -6.9329e+01   1e-01   9e-15   7e-15
 9: -6.9259e+01  -6.9261e+01   2e-03   2e-15   8e-15
10: -6.9260e+01  -6.9260e+01   2e-05   2e-15   7e-15
Optimal solution found.
{'beta': array([[ 7.54214409],
       [-1.65772882]]), 'beta_0': -0.14880733394172593}
```

9.4   Extension of Support Vector Machines Using a Kernel

The reason for solving the dual problem rather than the primary problem is that L_D can be expressed by the inner product $\langle \cdot, \cdot \rangle$ as

$$L_D := \sum_{i=1}^{N} \alpha_i - \frac{1}{2} \sum_{i=1}^{N} \sum_{j=1}^{N} \alpha_i \alpha_j y_i y_j \langle x_i, x_j \rangle .$$

Let V be the vector space with inner product $\phi : \mathbb{R}^p \to V$. Then, we may replace $\langle x_i, x_j \rangle$ by $k(x_i, x_j) := \langle \phi(x_i), \phi(x_j) \rangle$. In such a case, we construct a nonlinear classification rule from $(\phi(x_1), y_1), \ldots, (\phi(x_N), y_N)$. In other words, even if the mapping $\phi(x) \to y$ is linear and the learning via a support vector machine remains the same, the mapping $x \to y$ can be nonlinear. For the new data $(x_*, y_*) \in \mathbb{R}^p \times \{-1, 1\}$, the mapping $x_* \mapsto y_*$ is nonlinear.

In the following, let V be a vector space with an inner product; we construct a matrix K such that the (i, j)-th element is $\phi(x_i), \phi(x_j) \in V$. In fact, for $z \in \mathbb{R}^N$, the matrix

$$z^T K z = \sum_{i=1}^N \sum_{j=1}^N z_i \left\langle \phi(x_i), \phi(x_j) \right\rangle z_j$$

$$= \left\langle \sum_{i=1}^N z_i \phi(x_i), \sum_{j=1}^N z_j \phi(x_j) \right\rangle = \left\| \sum_{i=1}^N z_i \phi(x_i) \right\|^2 \geq 0$$

is symmetric and nonnegative definite, and $k(\cdot, \cdot)$ is a kernel in the strict sense.

Example 74 (Polynomial Kernel) For the d-dimensional polynomial kernel $k(x, y) = (1 + \langle x, y \rangle)^d$ with $x, y \in \mathbb{R}^p$, if $d = 1$ and $p = 2$, then since

$$1 + x_1 y_1 + x_2 y_2 = 1 \cdot 1 + x_1 y_1 + x_2 y_2 = \langle [1, x_1, x_2], [1, y_1, y_2] \rangle ,$$

we have $\phi : [x_1, x_2] \mapsto [1, x_1, x_2]$ with $V = \mathbb{R}^3$. For $p = 2$ and $d = 2$. Because

$$(1 + x_1 y_1 + x_2 y_2)^2 = 1 + x_1^2 y_1^2 + x_2^2 y_2^2 + 2x_1 y_1 + 2x_2 y_2 + 2x_1 x_2 y_1 y_2$$

$$= \langle [1, x_1^2, x_2^2, \sqrt{2} x_1, \sqrt{2} x_2, \sqrt{2} x_1 x_2],$$

$$[1, y_1^2, y_2^2, \sqrt{2} y_1, \sqrt{2} y_2, \sqrt{2} y_1 y_2] \rangle ,$$

we have

$$\phi : [x_1, x_2] \mapsto [1, x_1^2, x_2^2, \sqrt{2} x_1, \sqrt{2} x_2, \sqrt{2} x_1 x_2]$$

with $V = \mathbb{R}^6$. In this way, there exists $\phi : \mathbb{R}^p \to V$ such that $k(x, y) = \langle \phi(x), \phi(y) \rangle$. We can write the inner product and the polynomial kernel with $d = p = 2$ using the Python language as

```
def K_linear(x,y):
    return x.T@y
def K_poly(x,y):
    return (1+x.T@y)**2
```

We say that V is a vector space over $\mathbb{R}$ if

$$a, b \in V, \alpha, \beta \in \mathbb{R} \Longrightarrow \alpha a + \beta b \in V \qquad (9.24)$$

and that any element in V is a vector. There are various inner products. We say that the mapping $\langle \cdot, \cdot \rangle: V \times V \to \mathbb{R}$ is an inner product of V if the following properties hold for $a, b, c \in V, \alpha \in \mathbb{R}$ $\langle a + b, c \rangle = \langle a, c \rangle + \langle b, c \rangle$, $\langle a, b \rangle = \langle b, a \rangle$, $\langle \alpha a, b \rangle = \alpha \langle a, b \rangle$, $\langle a, a \rangle = \|a\|^2 \geq 0$, and $\|a\| = 0 \Longrightarrow a = 0$.

Example 75 The set V of continuous functions defined in $[0, 1]$ is a vector space. In fact, V satisfies (9.24). The function $V \times V \to \mathbb{R}$ defined by $\langle f, g \rangle :=$ $\int_0^1 f(x)g(x)dx$ for $f, g \in V$ is an inner product. In fact, we can check the four properties:

$$\langle f + g, h \rangle = \int_0^1 (f(x) + g(x))h(x)dx$$

$$= \int_0^1 f(x)h(x)dx + \int_0^1 g(x)h(x)dx = \langle f, h \rangle + \langle g, h \rangle$$

$$\langle f, g \rangle = \int_0^1 f(x)g(x)dx = \int_0^1 g(x)f(x)dx = \langle g, f \rangle$$

$$\langle \alpha f, g \rangle = \int_0^1 \alpha f(x) \cdot g(x)dx = \alpha \int_0^1 f(x)g(x)dx = \alpha \langle f, g \rangle$$

$$\langle f, f \rangle = \int_0^1 \{f(x)\}^2 dx \geq 0.$$

Moreover, since f is continuous, $\int_0^1 f^2(x)dx = 0$ if and only if $f = 0$.

Example 76 The map $V \times V \to \mathbb{R}$ defined by $f(x, y) := (1 + x^T y)^2, x, y \in \mathbb{R}^p$, is not an inner product of $V := \mathbb{R}^p$. In fact, since $f(0 \cdot x, y) = 1 \neq 0 \cdot f(x, y)$, the mapping does not satisfy the definition of an inner product.

In the following, we replace $x_i \in \mathbb{R}^p$, $i = 1, \ldots, N$, by $\phi(x_i) \in V$ using $\phi: \mathbb{R}^p \to V$. Therefore, $\beta \in \mathbb{R}^p$ is $\beta = \sum_{i=1}^N \alpha_i y_i \phi(x_i) \in V$, and the inner product $\langle x_i, x_j \rangle$ in L_D is replaced by the inner product of $\phi(x_i)$ and $\phi(x_j)$, the kernel of $K(x_i, x_j)$. If we extend in this way, the border becomes $\phi(X)\beta + \beta_0 = 0$, which means that $\sum_{i=1}^N \alpha_i y_i K(X, x_i) + \beta_0 = 0$ is not necessarily a surface.

We modify the function svm_1 as follows:

1. add argument K to the function definition,
2. replace np.dot(X[:,i]*X[:,j]) with K(X[i,:],X[j,:]), and
3. replace beta in return() with alpha.

In this manner, we can generalize the support vector machine.

```python
def svm_2(X,y,C,K):
    eps=0.0001
    n=X.shape[0]
    P=np.zeros((n,n))
    for i in range(n):
        for j in range(n):
            P[i,j]=K(X[i,:],X[j,:])*y[i]*y[j]
    # It must be specified using the matrix function in cvxopt
    P=matrix(P+np.eye(n)*eps)
    A=matrix(-y.T.astype(np.float))
    b=matrix(np.array([0]).astype(np.float))
    h=matrix(np.array([C]*n+[0]*n).reshape(-1,1).astype(np.float))
    G=matrix(np.concatenate([[np.diag(np.ones(n)),np.diag(-np.ones(n))]]))
    q=matrix(np.array([-1]*n).astype(np.float))

    res=cvxopt.solvers.qp(P,q,A=A,b=b,G=G,h=h)
    alpha=np.array(res['x'])   # where x corresponds to alpha in the text
    beta=((alpha*y).T@X).reshape(2,1)
    index=np.arange(0,n,1)
    index  (eps<alpha[:,0])&(alpha[:,0]<c-eps)
    beta_0=np.mean(y[index]-X[index,:]@beta)

    return {'alpha':alpha,'beta':beta,'beta_0':beta_0}
```

Example 77 Using the function svm_2, we compare the borders generated by linear and nonlinear kernels (Fig. 9.4).

```python
# execute
a=3;b=-1
n=200
X=randn(n,2)
y=np.sign(a*X[:,0]+b*X[:,1]**2+0.3*randn(n))
y=y.reshape(-1,1)
```

```python
def plot_kernel(K,line): # Specify the type of line by argument line.
    res=svm_2(X,y,1,K)
    alpha=res['alpha'][:,0]
    beta_0=res['beta_0']
    def f(u,v):
        S=beta_0
        for i in range(X.shape[0]):
            S=S+alpha[i]*y[i]*K(X[i,:],[u,v])
        return S[0]
    uu=np.arange(-2,2,0.1); vv=np.arange(-2,2,0.1); ww=[]
    for v in vv:
        w=[]
        for u in uu:
            w.append(f(u,v))
        ww.append(w)
    plt.contour(uu,vv,ww,levels=0,linestyles=line)
```

Fig. 9.4 Generating samples, we draw linear and nonlinear borders that are flat and curved surfaces, respectively

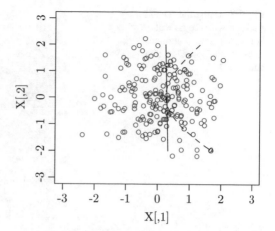

```
for i in range(n):
    if y[i]==1:
        plt.scatter(X[i,0],X[i,1],c="red")
    else:
        plt.scatter(X[i,0],X[i,1],c="blue")
plot_kernel(K_poly,line="dashed")
plot_kernel(K_linear,line="solid")
```

```
        pcost        dcost        gap    pres    dres
 0: -7.5078e+01  -6.3699e+02    4e+03   4e+00   3e-14
 1: -4.5382e+01  -4.5584e+02    9e+02   6e-01   2e-14
 2: -2.6761e+01  -1.7891e+02    2e+02   1e-01   1e-14
 3: -2.0491e+01  -4.9270e+01    4e+01   2e-02   1e-14
 4: -2.4760e+01  -3.3429e+01    1e+01   5e-03   5e-15
 5: -2.6284e+01  -2.9464e+01    4e+00   1e-03   3e-15
 6: -2.7150e+01  -2.7851e+01    7e-01   4e-05   4e-15
 7: -2.7434e+01  -2.7483e+01    5e-02   2e-06   5e-15
 8: -2.7456e+01  -2.7457e+01    5e-04   2e-08   5e-15
 9: -2.7457e+01  -2.7457e+01    5e-06   2e-10   6e-15
Optimal solution found.
        pcost        dcost        gap    pres    dres
 0: -9.3004e+01  -6.3759e+02    4e+03   4e+00   4e-15
 1: -5.7904e+01  -4.6085e+02    8e+02   5e-01   4e-15
 2: -3.9388e+01  -1.5480e+02    1e+02   6e-02   1e-14
 3: -4.5745e+01  -6.8758e+01    3e+01   9e-03   3e-15
 4: -5.0815e+01  -6.0482e+01    1e+01   3e-03   2e-15
 5: -5.2883e+01  -5.7262e+01    5e+00   1e-03   2e-15
 6: -5.3646e+01  -5.6045e+01    3e+00   6e-04   2e-15
 7: -5.4217e+01  -5.5140e+01    1e+00   2e-04   2e-15
 8: -5.4531e+01  -5.4723e+01    2e-01   1e-05   2e-15
 9: -5.4617e+01  -5.4622e+01    6e-03   3e-07   3e-15
10: -5.4619e+01  -5.4619e+01    6e-05   3e-09   3e-15
11: -5.4619e+01  -5.4619e+01    6e-07   3e-11   2e-15
Optimal solution found.
```

Thus far, we construct Python language programs to understand the principle. In actual data analysis, the svm in the sklearn package, is available for support vector machines.

Fig. 9.5 Using the e1071 package with the radical kernel, we draw a nonlinear (curved) surface. $C = 100$ and $\gamma = 1$

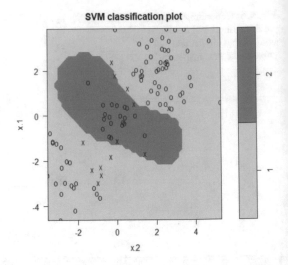

SVM classification plot

Example 78 For artificial data, using the svm in the sklearn package we executed the radical kernel

$$k(x, y) = \exp\left\{-\frac{1}{2\sigma^2}\|x - y\|^2\right\}$$

with $\gamma = 1$ for cost $C = 1$ (Fig. 9.5).

```
import sklearn
from sklearn import svm
```

```
x=randn(200,2)
x[0:100,]=x[0:100,]+2
x[100:150,]=x[100:150,]-2
y=np.concatenate((([1 for i in range(150)],[2 for i in range(50)])))
train=np.random.choice(200,100,replace=False)
test=list(set(range(200))-set(train))

res_svm=svm.SVC(kernel="rbf",gamma=1,C=100)   # SVM without tuning
res_svm.fit(x[train,],y[train])   # execute
```

```
SVC(C=100, cache_size=200, class_weight=None, coef0=0.0,
    decision_function_shape='ovr', degree=3, gamma=1, kernel='rbf', max_iter=-1,
    probability=False, random_state=None, shrinking=True, tol=0.001,
    verbose=False)
```

```
res_svm.predict(x[test,])        # prediction with test data
```

```
array([1, 1, 1, 1, 2, 1, 1, 2, 1, 2, 1, 2, 1, 1, 1, 1, 1, 1, 1, 1, 1, 1,
       1, 1, 2, 1, 1, 1, 1, 1, 1, 2, 1, 1, 1, 1, 1, 1, 1, 2, 1,
       1, 1, 1, 1, 1, 2, 1, 1, 2, 1, 1, 1, 1, 1, 1, 1, 1, 1, 1, 1, 1,
       1, 1, 2, 2, 2, 1, 2, 1, 2, 2, 2, 1, 2, 2, 2, 2, 1, 2, 2, 2, 2, 1,
       2, 2, 2, 2, 2, 1, 2, 2, 2, 1, 1, 1])
```

```
import mlxtend
from mlxtend.plotting import plot_decision_regions
```

```
plot_decision_regions(x,y,clf=res_svm)
```

Using the GridSearchCV command in the sklearn.model_selection, we compare via cross-validation the optimum combination of C and γ over $C = 0.1, 1, 10, 100, 1000$ and $\gamma = 0.5, 1, 2, 3, 4$ and find that the pair of $C = 1$ and $\gamma = 0.5$ is the best.

```
from sklearn.model_selection import GridSearchCV
```

```
grid={'C':[0.1,1,10,100,1000],'gamma':[0.5,1,2,3,4]}
tune=GridSearchCV(svm.SVC(),grid,cv=10)
tune.fit(x[train,],y[train])
```

```
GridSearchCV(cv=10, error_score='raise-deprecating',
             estimator=SVC(C=1.0, cache_size=200, class_weight=None, coef0=0.0,
                           decision_function_shape='ovr', degree=3,
                           gamma='auto_deprecated', kernel='rbf', max_iter=-1,
                           probability=False, random_state=None, shrinking=True,
                           tol=0.001, verbose=False),
             iid='warn', n_jobs=None,
             param_grid={'C': [0.1, 1, 10, 100, 1000],
                         'gamma': [0.5, 1, 2, 3, 4]},
             pre_dispatch='2*n_jobs', refit=True, return_train_score=False,
             scoring=None, verbose=0)
```

```
tune.best_params_    # we find that C=1 , gamma=0.5 are optimal.
```

```
{'C': 1, 'gamma': 0.5}
```

Example 79 In general, a support vector machine can execute even when the number of classes is more than two. For example, the function svm runs without specifying the number of classes. For Fisher's Iris dataset, we divide the 150 samples into 120 training and 30 test data to evaluate the performance: the kernel is radical and the parameters are $\gamma = 1$ and $C = 10$.

```
from sklearn.datasets import load_iris
```

```
iris=load_iris()
iris.target_names
x=iris.data
y=iris.target
train=np.random.choice(150,120,replace=False)
test=np.ones(150,dtype=bool)
test[train]=False
iris_svm=svm.SVC(kernel="rbf",gamma=1,C=10)
iris_svm.fit(x[train,],y[train])
```

```
SVC(C=10, cache_size=200, class_weight=None, coef0=0.0,
    decision_function_shape='ovr', degree=3, gamma=1, kernel='rbf', max_iter=-1,
    probability=False, random_state=None, shrinking=True, tol=0.001,
    verbose=False)
```

For example, we obtain the following result:

```
y_pre=iris_svm.predict(x[test,])
table_count(3,y[test],y_pre)
```

```
array([[ 9.,  0.,  0.],
       [ 0., 10.,  0.],
       [ 0.,  3.,  8.]])
```

Appendix: Proofs of Propositions

Proposition 23 *The distance between a point $(x, y) \in \mathbb{R}^2$ and a line $l : aX + bY + c = 0$, $a, b \in \mathbb{R}$ is given by*

$$\frac{|ax + by + c|}{\sqrt{a^2 + b^2}} .$$

Proof Let (x_0, y_0) be the perpendicular foot of l from (x, y). l' is a normal of l and can be written by

$$l' : \frac{X - x_0}{a} = \frac{Y - y_0}{b} = t$$

for some t (Fig. 9.6). Since (x_0, y_0) and (x, y) are on l and l', respectively, we have

$$\begin{cases} ax_0 + by_0 + c = 0 \\ \dfrac{x - x_0}{a} = \dfrac{y - y_0}{b} = t . \end{cases}$$

Fig. 9.6 The distance between a point and a line $\sqrt{(x - x_0)^2 + (y - y_0)^2}$, where l' is the normal of l that goes through (x_0, y_0)

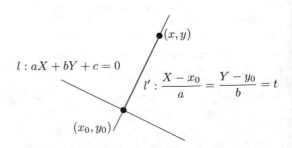

If we erase (x_0, y_0), from $x_0 = x - at$, $y_0 = y - bt$, $a(x - at) + b(y - bt) + c = 0$, we have $t = (ax + by + c)/(a^2 + b^2)$. Thus, the distance is

$$\sqrt{(x - x_0)^2 + (y - y_0)^2} = \sqrt{(a^2 + b^2)t^2} = \frac{|ax + by + c|}{\sqrt{a^2 + b^2}}.$$

□

Proposition 25

$$\begin{cases} \alpha_i = 0 & \Longleftarrow y_i(\beta_0 + x_i\beta) > 1 \\ 0 < \alpha_i < C \Longrightarrow y_i(\beta_0 + x_i\beta) = 1 \\ \alpha_i = C & \Longleftarrow y_i(\beta_0 + x_i\beta) < 1. \end{cases}$$

Proof When $\alpha_i = 0$, applying (9.20), (9.17), and (9.14) in this order, we have

$$\alpha_i = 0 \Longrightarrow \mu_i = C > 0 \Longrightarrow \epsilon_i = 0 \Longrightarrow y_i(\beta_0 + x_i\beta) \geq 1.$$

When $0 < \alpha_i < C$, from (9.17) and (9.20), we have $\epsilon_i = 0$. Moreover, applying (9.16), we have

$$0 < \alpha_i < C \Longrightarrow y_i(\beta_0 + x_i\beta) - (1 - \epsilon_i) = 0 \Longrightarrow y_i(\beta_0 + x_i\beta) = 1.$$

When $\alpha_i = C$, from (9.15), we have $\epsilon_i \geq 0$. Moreover, applying (9.16), we have

$$\alpha_i = C \Longrightarrow y_i(\beta_0 + x_i\beta) - (1 - \epsilon_i) = 0 \Longrightarrow y_i(\beta_0 + x_i\beta) \leq 1.$$

Furthermore, from (9.16), we have $y_i(\beta_0 + x_i\beta) > 1 \Longrightarrow \alpha_i = 0$. On the other hand, applying (9.14), (9.17), and (9.20) in this order, we have

$$y_i(\beta_0 + x_i\beta) < 1 \Longrightarrow \epsilon_i > 0 \Longrightarrow \mu_i = 0 \Longrightarrow \alpha_i = C.$$

□

Exercises 75–87

We define the distance between a point $(u, v) \in \mathbb{R}^2$ and a line $aU + bV + c = 0$, $a, b \in \mathbb{R}$ by

$$\frac{|au + bv + c|}{\sqrt{a^2 + b^2}}.$$

For $\beta \in \mathbb{R}^p$ such that $\beta_0 \in \mathbb{R}$ and $\|\beta\|_2 = 1$, when samples $(x_1, y_1), \ldots, (x_N, y_N) \in \mathbb{R}^p \times \{-1, 1\}$ satisfy the separability $y_1(\beta_0 + x_1\beta), \ldots, y_N(\beta_0 + x_N\beta) \geq 0$, the support vector machine is formulated as the problem of finding (β_0, β) that maximize the minimum value $M := \min_i y_i(\beta_0 + x_i\beta)$ over the distances between x_i (row vector) and the surface $\beta_0 + X\beta = 0$.

75. We extend the support vector machine problem to finding $(\beta_0, \beta) \in \mathbb{R} \times \mathbb{R}^p$ and $\epsilon_i \geq 0, i = 1, \ldots, N$, that maximize M under the constraints $\gamma \geq 0$, $M \geq 0$,
$$\sum_{i=1}^{N} \epsilon_i \leq \gamma, \text{ and }$$

$$y_i(\beta_0 + x_i\beta) \geq M(1 - \epsilon_i), \ i = 1, \ldots, N .$$

(a) What can we say about the locations of samples (x_i, y_i) when $\epsilon_i = 0$, $0 < \epsilon_i < 1$, $\epsilon_i = 1$, $1 < \epsilon_i$.
(b) Suppose that $y_i(\beta_0 + x_i\beta) < 0$ for at least r samples and for any β_0 and β. Show that if $\gamma \leq r$, then no solution exists. Hint: $\epsilon_i > 1$ for such an i.
(c) The larger the γ is, the smaller the M. Why?

76. We wish to obtain $\beta \in \mathbb{R}^p$ that minimizes $f_0(\beta)$ under $f_j(\beta) \leq 0, j = 1, \ldots, m$. If such a solution exists, we denote the minimum value by f^*. Consider the following two equations

$$\sup_{\alpha \geq 0} L(\alpha, \beta) = \begin{cases} f_0(\beta), & f_j(\beta) \leq 0 , \ j = 1, \ldots, m \\ +\infty & \text{Otherwise} \end{cases} \qquad (9.25)$$

$$f^* := \inf_{\beta} \sup_{\alpha \geq 0} L(\alpha, \beta) \geq \sup_{\alpha \geq 0} \inf_{\beta} L(\alpha, \beta) \qquad (9.26)$$

under

$$L(\alpha, \beta) := f_0(\beta) + \sum_{j=1}^{m} \alpha_j f_j(\beta)$$

for $\alpha = (\alpha_1, \ldots, \alpha_m) \in \mathbb{R}^m$. Moreover, suppose $p = 2$ and $m = 1$. For

$$L(\alpha, \beta) := \beta_1 + \beta_2 + \alpha(\beta_1^2 + \beta_2^2 - 1) , \qquad (9.27)$$

such that the equality holds in the inequality (9.26).

77. Suppose that $f_0, f_1, \ldots, f_m : \mathbb{R}^p \to \mathbb{R}$ are convex and differentiable at $\beta = \beta^*$. It is known that $\beta^* \in \mathbb{R}^p$ is the optimum value of $\min\{f_0(\beta) \mid f_i(\beta) \leq 0, i = 1, \ldots, m\}$ if and only if there exist $\alpha_i \geq 0, i = 1, \ldots, m$, such that

$$f_i(\beta^*) \leq 0, \ i = 1, \ldots, m, \qquad (9.28)$$

and the two conditions are met (KKT conditions)

$$\alpha_i f_i(\beta^*) = 0, \quad i = 1, \ldots, m, \tag{9.29}$$

$$\nabla f_0(\beta^*) + \sum_{i=1}^{m} \alpha_i \nabla f_i(\beta^*) = 0. \tag{9.30}$$

In this problem, we consider the sufficiency.

(a) If $f : \mathbb{R}^p \to \mathbb{R}$ is convex and differentiable at $x = x_0 \in \mathbb{R}$, then

$$f(x) \geq f(x_0) + \nabla f(x_0)^T (x - x_0) \tag{9.31}$$

for each $x \in \mathbb{R}^p$. From this fact, show that $f_0(\beta^*) \leq f_0(\beta)$ for arbitrary $\beta \in \mathbb{R}^p$ that satisfies (9.28). Hint: Use (9.29) and (9.30) once, (9.31) twice, and $f_1(\beta) \leq 0, \ldots, f_m(\beta) \leq 0$ once.

(b) For (9.27), find the conditions that correspond to (9.28)–(9.30).

78. If we remove the condition $\|\beta\|_2 = 1$ in Problem 75 and regard $\beta_0/M, \beta/M$ as β_0 and β, then the problem reduces to finding $\beta_0, \beta, \epsilon_i, i = 1, \ldots, N$, that minimize

$$L_P := \frac{1}{2}\|\beta\|_2^2 + C \sum_{i=1}^{N} \epsilon_i - \sum_{i=1}^{N} \alpha_i \{y_i(\beta_0 + x_i\beta) - (1 - \epsilon_i)\} - \sum_{i=1}^{N} \mu_i \epsilon_i, \tag{9.32}$$

where $C > 0$ (cost), the last two terms are constraints, and $\alpha_i, \mu_i \geq 0, i = 1, \ldots, N$, are the Lagrange coefficients. Show that the KKT conditions (9.28)–(9.30) are the following:

$$\sum_{i=1}^{N} \alpha_i y_i = 0 \tag{9.33}$$

$$\beta = \sum_{i=1}^{N} \alpha_i y_i x_i \in \mathbb{R}^p \tag{9.34}$$

$$C - \alpha_i - \mu_i = 0 \tag{9.35}$$

$$\alpha_i [y_i(\beta_0 + x_i\beta) - (1 - \epsilon_i)] = 0 \tag{9.36}$$

$$\mu_i \epsilon_i = 0 \tag{9.37}$$

$$y_i(\beta_0 + x_i\beta) - (1 - \epsilon_i) \geq 0 \tag{9.38}$$

$$\epsilon_i \geq 0. \tag{9.39}$$

79. Show that the dual problem (9.32) of L_P is given by

$$L_D := \sum_{i=1}^{N} \alpha_i - \frac{1}{2} \sum_{i=1}^{N} \sum_{j=1}^{N} \alpha_i \alpha_j y_i y_j x_i^T x_j , \tag{9.40}$$

where α ranges over (9.33) and

$$0 \le \alpha_i \le C . \tag{9.41}$$

Moreover, how is β obtained from such an α?

80. Show the following:

$$\begin{cases} \alpha_i = 0 & \Longleftarrow y_i(\beta_0 + x_i\beta) > 1 \\ 0 < \alpha_i < C \Longrightarrow y_i(\beta_0 + x_i\beta) = 1 \\ \alpha_i = C & \Longleftarrow y_i(\beta_0 + x_i\beta) < 1. \end{cases}$$

81. We wish to obtain the value of β_0 by $y_i(\beta_0 + x_i\beta) = 1$ for at least one i.

   (a) Show that $\alpha_1 = \cdots = \alpha_N = 0$ and $y_i(\beta_0 + x_i\beta) = 1$ imply $\beta_0 = y_i$, $i = 1, \ldots, N$.
   (b) Suppose that ($\alpha = 0$ or $\alpha = C$) and $y_i(\beta_0 + x_i\beta) \neq 1$ for each i, and let $\epsilon_* := \min_i \epsilon_i$. Show that L_p decreases when replacing ϵ_i and β by $\epsilon_i - \epsilon_*$ and $\beta_0 + y_i\epsilon_*$, respectively, for each i, which means that no optimum solution can be obtained under the assumption. Hint: $y_i = \pm 1 \Longleftrightarrow y_i^2 = 1$.
   (c) Show that $y_i(\beta_0 + x_i\beta) = 1$ for at least one i.

82. In order to input the dual problem (9.40), (9.33), and (9.41) into a quadratic programming solver, we specify $D_{mat} \in \mathbb{R}^{N \times N}$, $A_{mat} \in \mathbb{R}^{m \times N}$, $d_{vec} \in \mathbb{R}^N$, and $b_{vec} \in \mathbb{R}^m$ ($m \ge 1$) such that

$$L_D = -\frac{1}{2}\alpha^T D_{mat}\alpha + d_{vec}^T\alpha$$

$$A_{mat}\alpha \ge b_{vec} ,$$

where the first meq and the last $m - meq$ are equalities and inequalities, respectively, in the m constraints $A_{mat}\alpha \ge b_{vec}$, $\alpha \in \mathbb{R}^N$. If we define

$$b_{vec} := [0, -C, \ldots, -C, 0, \ldots, 0]^T ,$$

what are D_{mat}, A_{mat}, d_{vec}, and meq? Moreover, fill in the blanks below and execute the result.

```
import cvxopt
from cvxopt import matrix
```

```
a=randn(1); b=randn(1)
n=100
X=randn(n,2)
y=np.sign(a*X[:,0]+b*X[:,1]+0.1*randn(n))
y=y.reshape(-1,1)   # The shape needs to be clearly marked
```

```
def svm_1(X,y,C):
    eps=0.0001
    n=X.shape[0]
    P=np.zeros((n,n))
    for i in range(n):
        for j in range(n):
            P[i,j]=np.dot(X[i,:],X[j,:])*y[i]*y[j]
    # It must be specified using the matrix function in cvxopt.
    P=matrix(P+np.eye(n)*eps)
    A=matrix(-y.T.astype(np.float))
    b=matrix(# blank(1) #).astype(np.float))
    h=matrix(# blank(2) #).reshape(-1,1).astype(np.float))
    G=matrix(np.concatenate([# blank(3) #,np.diag(-np.ones(n))]))
    q=matrix(np.array([-1]*n).astype(np.float))
    res=cvxopt.solvers.qp(P,q,A=A,b=b,G=G,h=h)      # execute solver
    alpha=np.array(res['x'])   # # where x corresponds to alpha in the
        text
    beta=((alpha*y).T@X).reshape(2,1)
    index=np.arange(0,n,1)
    index_1=index[eps<alpha[:,0]]
    index_2=index[(alpha<C-eps)[:,0]]
    index=np.concatenate((index_1,index_2))
    beta_0=np.mean(y[index]-X[index,:]@beta)
    return {'beta':beta,'beta_0':beta_0}
```

```
a=randn(1); b=randn(1)
n=100
X=randn(n,2)
y=np.sign(a*X[:,0]+b*X[:,1]+0.1*randn(n))
y=y.reshape(-1,1)   # The shape needs to be clearly marked
for i in range(n):
    if y[i]==1:
        plt.scatter(X[i,0],X[i,1],c="red")
    else :
        plt.scatter(X[i,0],X[i,1],c="blue")
res=svm_1(X,y,C=10)
```

```
def f(x):
    return -res['beta_0']/res['beta'][1]-x*res['beta'][0]/res['beta'][1]
```

```
x_seq=np.arange(-3,3,0.5)
plt.plot(x_seq,f(x_seq))
```

83. Let V be a vector space. We define a kernel $K(x, y)$ w.r.t. $\phi : \mathbb{R}^p \to V$ as the inner product of $\phi(x)$ and $\phi(y)$ given $(x, y) \in \mathbb{R}^p \times \mathbb{R}^p$. For example, for the

d-dimensional polynomial kernel $K(x, y) = (1 + x^T y)^d$, if $d = 1$ and $p = 2$, then the mapping is

$$((x_1, x_2), (y_1, y_2)) \mapsto 1 \cdot 1 + x_1 y_1 + x_2 y_2 = (1, x_1, x_2)^T (1, y_1, y_2) \, .$$

In this case, we regard the map ϕ as $(x_1, x_2) \mapsto (1, x_1, x_2)$. What is ϕ for $p = 2$ and $d = 2$? Write a Python function K_poly(x,y) that realizes the $d = 2$-dimensional polynomial kernel.

84. Let V be a vector space over $\mathbb{R}$.

   (a) Suppose that V is the set of continuous functions in $[0, 1]$. Show that
   $$\int_0^1 f(x)g(x)dx, \ f, g \in V, \text{ is an inner product of } V.$$
   (b) For vector space $V := \mathbb{R}^p$, show that $(1 + x^T y)^2$, $x, y \in \mathbb{R}^p$, is not an inner product of V.
   (c) Write a Python function K_linear(x,y) for the standard inner product.

   Hint: Check the definition of an inner product, for $a, b, c \in V$, $\alpha \in \mathbb{R}$, $\langle a + b, c \rangle = \langle a, c \rangle + \langle b, c \rangle$; $\langle a, b \rangle = \langle b, a \rangle$; $\langle \alpha a, b \rangle = \alpha \langle a, b \rangle$; $\langle a, a \rangle = \|a\|^2 \geq 0$.

85. In the following, using $\phi : \mathbb{R}^p \to V$, we replace $x_i \in \mathbb{R}^p$, $i = 1, \ldots, N$, with $\phi(x_i) \in V$. Thus, $\beta \in \mathbb{R}^p$ is expressed as $\beta = \sum_{i=1}^{N} \alpha_i y_i \phi(x_i) \in V$, and the inner product $\langle x_i, x_j \rangle$ in L_D is replaced by the inner product of $\phi(x_i)$ and $\phi(x_j)$, i.e., $K(x_i, x_j)$. If we extend the vector space, the border $\phi(X)\beta + \beta_0 = 0$, i.e., $\sum_{i=1}^{N} \alpha_i y_i K(X, x_i) + \beta_0 = 0$, is not necessarily a surface. Modify the svm_1 in Problem 82 as follows:

   (a) add argument K to the definition,
   (b) replace np.dot(X[,i]*X[,j]) with K(X[i,],X[j,]), and
   (c) replace beta in return by alpha.

   Then, execute the function svm_2 by filling in the blanks.

```
# generating data
a=3;b=-1
n=200
X=randn(n,2)
y=np.sign(a*X[:,0]+b*X[:,1]**2+0.3*randn(n))
y=y.reshape(-1,1)
```

```
def plot_kernel(K,line): # Specify the type of line by argument line
    res=svm_2(X,y,1,K)
    alpha=res['alpha'][:,0]
    beta_0=res['beta_0']
    def f(u,v):
        S=beta_0
        for i in range(X.shape[0]):
            S=S+ # blank #
        return S[0]
    uu=np.arange(-2,2,0.1); vv=np.arange(-2,2,0.1); ww=[]
    for v in vv:
        w=[]
```

```
        for u in uu:
            w.append(f(u,v))
        ww.append(w)
    plt.contour(uu,vv,ww,levels=0,linestyles=line)
```

```
for i in range(n):
    if y[i]==1:
        plt.scatter(X[i,0],X[i,1],c="red")
    else:
        plt.scatter(X[i,0],X[i,1],c="blue")
plot_kernel(K_poly,line="dashed")
plot_kernel(K_linear,line="solid")
```

```
       pcost          dcost        gap      pres     dres
 0: -7.5078e+01  -6.3699e+02   4e+03    4e+00    3e-14
 1: -4.5382e+01  -4.5584e+02   9e+02    6e-01    2e-14
 2: -2.6761e+01  -1.7891e+02   2e+02    1e-01    1e-14
 3: -2.0491e+01  -4.9270e+01   4e+01    2e-02    1e-14
 4: -2.4760e+01  -3.3429e+01   1e+01    5e-03    5e-15
 5: -2.6284e+01  -2.9464e+01   4e+00    1e-03    3e-15
 6: -2.7150e+01  -2.7851e+01   7e-01    4e-05    4e-15
 7: -2.7434e+01  -2.7483e+01   5e-02    2e-06    5e-15
 8: -2.7456e+01  -2.7457e+01   5e-04    2e-08    5e-15
 9: -2.7457e+01  -2.7457e+01   5e-06    2e-10    6e-15
Optimal solution found.
       pcost          dcost        gap      pres     dres
 0: -9.3004e+01  -6.3759e+02   4e+03    4e+00    4e-15
 1: -5.7904e+01  -4.6085e+02   8e+02    5e-01    4e-15
 2: -3.9388e+01  -1.5480e+02   1e+02    6e-02    1e-14
 3: -4.5745e+01  -6.8758e+01   3e+01    9e-03    3e-15
 4: -5.0815e+01  -6.0482e+01   1e+01    3e-03    2e-15
 5: -5.2883e+01  -5.7262e+01   5e+00    1e-03    2e-15
 6: -5.3646e+01  -5.6045e+01   3e+00    6e-04    2e-15
 7: -5.4217e+01  -5.5140e+01   1e+00    2e-04    2e-15
 8: -5.4531e+01  -5.4723e+01   2e-01    1e-05    2e-15
 9: -5.4617e+01  -5.4622e+01   6e-03    3e-07    3e-15
10: -5.4619e+01  -5.4619e+01   6e-05    3e-09    3e-15
11: -5.4619e+01  -5.4619e+01   6e-07    3e-11    2e-15
Optimal solution found.
```

(a) Execute the support vector machine with $\gamma = 1$ and $C = 100$.

```
import sklearn
from sklearn import svm
```

```
x=randn(200,2)
x[0:100,]=x[0:100,]+2
x[100:150,]=x[100:150,]-2
y=np.concatenate(([1 for i in range(150)],[2 for i in range(50)]))
train=np.random.choice(200,100,replace=False)
test=list(set(range(200))-set(train))
res_svm=svm.SVC(kernel="rbf",gamma=1,C=1)  # SVM without tuning
res_svm.fit(x[train,],y[train])  # execute
```

```
res_svm.predict(x[test,])         # prediction with test data
```

```
import mlxtend
from mlxtend.plotting import plot_decision_regions
```

```
plot_decision_regions(x,y,clf=res_svm)
```

(b) Use the `GridSearchCV` cosmmand to find the optimal C and γ over $C = 0.1, 1, 10, 100, 1000$ and $\gamma = 0.5, 1, 2, 3, 4$ via cross-validation.

```
from sklearn.model_selection import GridSearchCV
```

```
grid={'C':[0.1,1,10,100,1000],'gamma':[0.5,1,2,3,4]}
tune=GridSearchCV(svm.SVC(),grid,cv=10)
tune.fit(x[train,],y[train])
```

```
GridSearchCV(cv=10, error_score='raise-deprecating',
             estimator=SVC(C=1.0, cache_size=200, class_weight=None, coef0=0.0,
                           decision_function_shape='ovr', degree=3,
                           gamma='auto_deprecated', kernel='rbf', max_iter=-1,
                           probability=False, random_state=None, shrinking=True,
                           tol=0.001, verbose=False),
             iid='warn', n_jobs=None,
             param_grid={'C': [0.1, 1, 10, 100, 1000],
                         'gamma': [0.5, 1, 2, 3, 4]},
             pre_dispatch='2*n_jobs', refit=True, return_train_score=False,
             scoring=None, verbose=0)
```

86. A support vector machine works even when more than two classes exist. In fact, the function `svm` in the `sklearn` packages runs even if we give no information about the number of classes. Fill in the blanks and execute it.

```
from sklearn.datasets import load_iris
```

```
iris=load_iris()
iris.target_names
x=iris.data
y=iris.target
train=np.random.choice(150,120,replace=False)
test=np.ones(150,dtype=bool)
test[train]=False
iris_svm=svm.SVC(kernel="rbf",gamma=1,C=10)
iris_svm.fit(# blank(1) #)
```

```
SVC(C=10, cache_size=200, class_weight=None, coef0=0.0,
    decision_function_shape='ovr', degree=3, gamma=1, kernel='rbf', max_iter=-1,
    probability=False, random_state=None, shrinking=True, tol=0.001,
    verbose=False)
```

```
y_pre=# blank(2) #
table_count(3,y[test],y_pre)
```

```
array([[ 9.,  0.,  0.],
       [ 0., 10.,  0.],
       [ 0.,  3.,  8.]])
```

Chapter 10
Unsupervised Learning

Abstract Thus far, we have considered supervised learning from N observation data $(x_1, y_1), \ldots, (x_N, y_N)$, where $y_1, \ldots, y_N$ take either real values (regression) or a finite number of values (classification). In this chapter, we consider unsupervised learning, in which such a teacher does not exist, and the relations between the N samples and between the p variables are learned only from covariates $x_1, \ldots, x_N$. There are various types of unsupervised learning; in this chapter, we focus on clustering and principal component analysis. Clustering means dividing the samples $x_1, \ldots, x_N$ into several groups (clusters). We consider K-means clustering, which requires us to give the number of clusters K in advance, and hierarchical clustering, which does not need such information. We also consider the principal component analysis (PCA), a data analysis method that is often used for machine learning and multivariate analysis. For PCA, we consider another equivalent definition along with its mathematical meaning.

10.1 K-means Clustering

Clustering divides N samples $x_1, \ldots, x_N$ with p variable values into K disjoint sets (clusters). Among the clustering methods, K-means clustering requires K to be determined in advance. In the initial stage, one of $1, \ldots, K$ is randomly assigned to each of the N samples, and we execute the following two steps:

1. for each cluster $k = 1, \ldots, K$, find the center (mean vector), and
2. for each sample $i = 1, \cdots, N$, assign the cluster such that the center is the closest among the K clusters,

---

The original version of this chapter was revised: typo error 'principle component' and 'principle component analysis' have been changed to 'principal component' and 'principal component analysis' respectively. The correction to this chapter can be found at https://doi.org/10.1007/978-981-15-7877-9_11

J. Suzuki, *Statistical Learning with Math and Python*, https://doi.org/10.1007/978-981-15-7877-9_10

where a cluster is a set of p-dimensional vectors and its (arithmetic) mean corresponds to the center. In the second step, we evaluate the distance in terms of the L_2 norm

$$\|a - b\| = \sqrt{(a_1 - b_1)^2 + \cdots + (a_p - b_p)^2}$$

for $a = [a_1, \ldots, a_p]^T$, $b = [b_1, \ldots, b_p]^T \in \mathbb{R}^p$.

For example, the following procedure can be constructed. If a cluster contains no samples in the middle of the run, the cluster is no longer used since its center cannot be calculated. This can occur if N is relatively small compared to K. In addition, in the following code, the line with # is not for generating the clusters but for tracing the changes in the score values.

```
def k_means(X,K,iteration=20):
    n,p=X.shape
    center=np.zeros((K,p))
    y=np.random.choice(K,n,replace=True)
    scores=[]
    for h in range(iteration):
        for k in range(K):
            if np.sum(y==k)==0:
                center[k,0]=np.inf
            else:
                for j in range(p):
                    center[k,j]=np.mean(X[y==k,j])
        S_total=0
        for i in range(n):
            S_min=np.inf
            for k in range(K):
                S=np.sum((X[i,]-center[k,])**2)
                if S<S_min:
                    S_min=S
                    y[i]=k
            S_total+=S_min
        scores.append(S_total)
    return {'clusters':y,'scores':scores}
```

Example 80 K-means clustering executed with the function k_means to display clusters of $p = 2$-dimensional artificial data (Fig. 10.1).

```
n=1000; K=5; p=2
X=randn(n,p)   # data generation
y=k_means(X,5)['clusters'] # getting cluster for each sample
#   Change the color of each cluster and draw a dot
plt.scatter(X[:,0],X[:,1],c=y)
plt.xlabel("first_component")
plt.ylabel("second_component")
```

```
Text(0, 0.5, 'second component')
```

Fig. 10.1 K-means
clustering with $K = 5$,
$N = 1000$, and $p = 2$.
Samples in the same cluster
share the same color

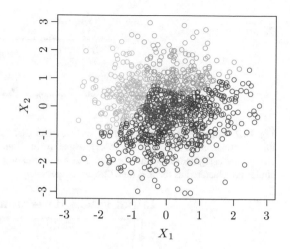

We see that the score

$$S := \sum_{k=1}^{K} \min_{z_k \in \mathbb{R}^p} \sum_{i \in C_k} \|x_i - z_k\|^2$$

does not increase for each update of Steps 1 and 2 while executing K-means
clustering, where C_k is the set of indexes i of samples in the kth cluster. In fact,
the square sum $\sum_{i \in C_k} \|x_i - x\|^2$ of the distances between $x \in \mathbb{R}^p$ and the points is
minimized when x is chosen to be the center $\bar{x}_k$ of cluster k:

$$\sum_{i \in C_k} \|x_i - x\|^2 = \sum_{i \in C_k} \|(x_i - \bar{x}_k) - (x - \bar{x}_k)\|^2$$

$$= \sum_{i \in C_k} \|x_i - \bar{x}_k\|^2 + \sum_{i \in C_k} \|x - \bar{x}_k\|^2 - 2(x - \bar{x}_k)^T \sum_{i \in C_k} (x_i - \bar{x}_k)$$

$$= \sum_{i \in C_k} \|x_i - \bar{x}_k\|^2 + \sum_{i \in C_k} \|x - \bar{x}_k\|^2 \geq \sum_{i \in C_k} \|x_i - \bar{x}_k\|^2 ,$$

where we have used $\bar{x}_k = \dfrac{1}{|C_k|} \sum_{i \in C_k} x_i$. In other words, the score does not increase
after executing Step 1. Moreover, in Step 2, if the cluster to which a sample belongs
is changed to the nearest cluster, the score S does not increase either.

Moreover, the result of K-means clustering depends on the randomly selected
initial clusters, which means that even if K-means clustering is applied, there is no
guarantee that an optimum solution will be obtained, and it is necessary to execute
it several times with different initial values and select the clustering with the best
score.

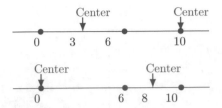

Fig. 10.2 Suppose that we have three samples and that the left two and right one circles are initially clustered as red and blue (upper), respectively. Then, K-means clustering does not further change the assignments of the clusters. The same situation occurs if the left one and right two circles are clustered as red and blue (lower), respectively

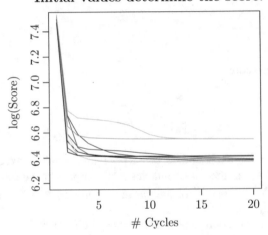

Fig. 10.3 K-means clustering repeated 10 times changing the initial values: each run uses a different color. The score decreases monotonically with each update, and the values at convergence do not match in each update. The horizontal and vertical axes express the number of iterations and the logarithm of the score, respectively

Example 81 Suppose that there are samples at $0, 6, 10$ with $N = 3$, $p = 1$, and $K = 2$ (Fig. 10.2). If "red" and "blue" are initially assigned to $0, 6$ and 10, respectively, then 3 and 8 are the centers of the clusters. In that case, the centers of the nearest clusters of $0, 6, 10$ are $3, 3, 10$ (the square error is $9 + 9 + 0 = 18$), and the clusters do not change even if we continue the process. Conversely, if "red" and "blue" are initially assigned to 0 and 6, 10, respectively, then 0 and 8 are the centers of the clusters. In that case, the closest cluster centers of $0, 6, 10$ are $0, 8, 8$ (the square error is $0 + 4 + 4 = 8$), and the clusters do not change even if we continue the process. The latter is optimal in the sense of the minimum score, but once the first configuration appears, we do not reach the optimal solution.

Example 82 We changed the initial values and repeated K-means clustering (Fig. 10.3). The score decreased in each execution, and the converged value differed for each initial value, which means that some executions have not reached the optimum. The code was as follows:

```
n=1000; p=2
X=randn(n,p)
itr=np.arange(1,21,1)
for r in range(10):
    scores=k_means(X,5)['scores']
    plt.plot(itr,np.log(scores))
plt.xlabel("␣the␣number␣of␣iteration")
plt.ylabel("log(value)")
plt.title("See␣how␣the␣score␣changes␣with␣each␣initial␣value")
plt.xticks(np.arange(1,21,1))
```

Let $x_1 = [x_{1,1}, \ldots, x_{1,p}]^T, \ldots, x_N = [x_{N,1}, \ldots, x_{N,p}]^T$ be N samples. If we write the set of sample indexes that belong to a cluster k (a subset of $\{1, \ldots, N\}$) and its center as C_k and $\bar{x}_k = [\bar{x}_{k,1}, \ldots, \bar{x}_{k,p}]^T$, respectively, for each $k = 1, \ldots, K$, we have the following relation:

$$\frac{1}{|C_k|} \sum_{i,i' \in C_k} \sum_{j=1}^{p} (x_{i,j} - x_{i',j})^2 = 2 \sum_{i \in C_k} \sum_{j=1}^{p} (x_{i,j} - \bar{x}_{k,j})^2 \tag{10.1}$$

In fact, (10.1) holds if

$$\frac{1}{|C_k|} \sum_{i \in C_k} \sum_{i' \in C_k} (x_{i,j} - x_{i',j})^2 = 2 \sum_{i \in C_k} (x_{i,j} - \bar{x}_{k,j})^2.$$

for $j = 1, \ldots, p$. In particular, the left-hand side can be transformed into

$$\frac{1}{|C_k|} \sum_{i \in C_k} \sum_{i' \in C_k} \{(x_{i,j} - \bar{x}_{k,j}) - (x_{i',j} - \bar{x}_{k,j})\}^2$$

$$= \frac{1}{|C_k|} \sum_{i \in C_k} \sum_{i' \in C_k} (x_{i,j} - \bar{x}_{k,j})^2 - \frac{2}{|C_k|} \sum_{i \in C_k} (x_{i,j} - \bar{x}_{k,j}) \sum_{i' \in C_k} (x_{i',j} - x_{k,j})$$

$$+ \frac{1}{|C_k|} \sum_{i \in C_k} \sum_{i' \in C_k} (x_{i',j} - \bar{x}_{k,j})^2, \tag{10.2}$$

where the second term of (10.2) is zero due to $\bar{x}_{k,j} = \dfrac{1}{|C_k|} \sum_{i' \in C_k} x_{i',j}$, the first and third terms in (10.2) share the same value $\sum_{i \in C_k} (x_{i,j} - \bar{x}_{k,j})^2$, and its sum coincides with the right-hand side of (10.1).

From (10.1), we see that K-means clustering seeks the configuration that minimizes the squared sum of the distances of the sample pairs in the clusters.

10.2   Hierarchical Clustering

Hierarchical clustering is another commonly used clustering method.

Initially, we construct N clusters, each of which contains only one sample. Then, based on a criterion, we merge two clusters in each stage until the number of clusters is two. For each of the stages $k = N, N - 1, \cdots, 2$, one clustering exists. The clusters for $k = N, N - 1, \cdots, 2$ can be obtained by returning the history of connections without determining the number of clusters K.

We use the $L2$ norm for the distance $d(\cdot, \cdot)$ between samples. However, it is necessary to define the distance between clusters that contain multiple samples (which does not necessarily satisfy the axiom of distance). The frequently used definitions are listed in Table 10.1.

The following procedure can be implemented for each of the complete, single, centroid, and average linkages. However, the input is given by a pair of matrices x, y that extract multiple rows of $X \in \mathbb{R}^{N \times p}$. The distance between the clusters is the output.

```
def dist_complete(x,y):
    r=x.shape[0]
    s=y.shape[0]
    dist_max=0
    for i in range(r):
        for j in range(s):
            d=np.linalg.norm(x[i,]-y[j,])**2
            if d>dist_max:
                dist_max=d
    return dist_max
```

```
def dist_single(x,y):
    r=x.shape[0]
    s=y.shape[0]
    dist_min=np.inf
    for i in range(r):
        for j in range(s):
            d=np.linalg.norm(x[i,]-y[j,])**2
            if d<dist_min:
                dist_min=d
    return dist_min
```

Table 10.1 Distance between clusters

Linkage	Definition	Distance between clusters A and B
Complete	The maximum distance between the clusters	$\max\limits_{i \in A, j \in B} d(x_i, x_j)$
Single	The minimum distance between the clusters	$\min\limits_{i \in A, j \in B} d(x_i, x_j)$
Centroid	The distance between the centers of the clusters	$d\left(\dfrac{1}{\|A\|} \sum\limits_{i \in A} x_i, \dfrac{1}{\|B\|} \sum\limits_{j \in B} x_j \right)$
Average	The mean distance between the clusters	$\dfrac{1}{\|A\| \cdot \|B\|} \sum\limits_{i \in A} \sum\limits_{j \in B} d(x_i, x_j)$

```
def dist_centroid(x,y):
    r=x.shape[0]
    s=y.shape[0]
    x_bar=0
    for i in range(r):
        x_bar=x_bar+x[i,]
    x_bar=x_bar/r
    y_bar=0
    for i in range(s):
        y_bar=y_bar+y[i,]
    y_bar=y_bar/s
    return (np.linalg.norm(x_bar-y_bar)**2)
```

```
def dist_average(x,y):
    r=x.shape[0]
    s=y.shape[0]
    S=0
    for i in range(r):
        for j in range(s):
            S=S+np.linalg.norm(x[i,]-y[j,])**2
    return (S/r/s)
```

Furthermore, when the distance between such clusters is defined, the procedure of hierarchical clustering can be defined as follows. Given the distances between samples and between clusters, the clustering is obtained (a list called index), and the list that consists of such lists is called cluster. If two clusters $i < j$ are connected, then cluster j is absorbed into cluster i, and j disappears. The indices $j + 1$ or larger are decreased by one, and the cluster k is deleted.

```
import copy
```

```
def hc(X,dd="complete"):
    n=X.shape[0]
    index=[[i] for i in range(n)]
    cluster=[[] for i in range(n-1)]
    for k in range(n,1,-1):
        # index_2=[]
        dist_min=np.inf
        for i in range(k-1):
            for j in range(i+1,k):
                i_0=index[i]; j_0=index[j]
                if dd=="complete":
                    d=dist_complete(X[i_0,],X[j_0,])
                elif dd=="single":
                    d=dist_single(X[i_0,],X[j_0,])
                elif dd=="centroid":
                    d=dist_centroid(X[i_0,],X[j_0,])
                elif dd=="average":
                    d=dist_average(X[i_0,],X[j_0,])
                if d<dist_min:
                    dist_min=d
                    i_1=i    # list of index which would be combined
                    j_1=j    # list of index which would join
        index[i_1].extend(index[j_1])  # add
        if j_1<k:  # put the added index forward
```

```
        for h in range(j_1+1,k,1):
            index[h-1]=index[h]
    index2=copy.deepcopy(index[0:(k-1)])    # If you use "index" without
        deepcopy , "index" will be rewritten each time.
    cluster[k-2].extend(index2)
    return cluster # The results from below show that one by one, the
    congruence occurs.
```

Thus, clusterings of sizes $k = n, n - 1, \cdots, 2$ are stored in `cluster[[n]]`, `cluster[[n-1]]`,...,`cluster[[2]]`.

Example 83 Hierarchical clustering was performed for artificially generated data with $N = 100$ and $p = 2$. Samples in the same cluster are shown in the same color. First, we changed the number of clusters K and output the results (Fig. 10.4).

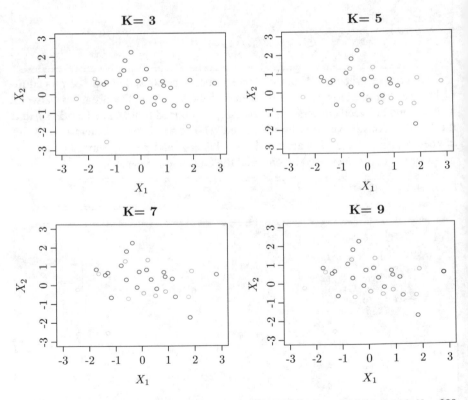

Fig. 10.4 We execute hierarchical clustering (complete linkage) for artificial data with $N = 200$, $p = 2$ and $K = 3, 5, 7, 9$. The samples in the same cluster share the same color. Compared to k-means clustering, not all the samples belong to the cluster with the nearest center

```
n=200; p=2
X=randn(n,p)
cluster=hc(X,"complete")
K=[2,4,6,8] # the number of clusters are 3,5,7 and 9.
for i in range(4):
    grp=cluster[K[i]]  # From the overall result, the result for K[i] is
        taken
    plt.subplot(2,2,i+1)
    for k in range(len(grp)):
        x=X[grp[k],0]
        y=X[grp[k],1]
        plt.scatter(x,y,s=5)
    plt.text(2,2,"K={}".format(K[i]+1),fontsize=12)
```

Next, we changed the definition of the distance between clusters (complete, single, average, and centroid) and output the results, as shown in Fig. 10.5. Samples in the same cluster are shown in the same color.

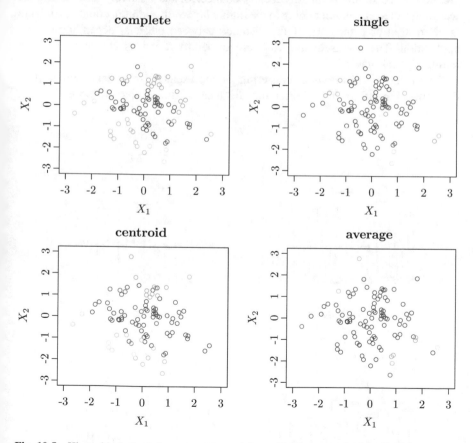

Fig. 10.5 Hierarchical clustering was performed for artificial data with $N = 100$, $p = 2$, and $K = 7$ for each complete, single, centroid, and average linkages. The most commonly used complete linkage appears to result in intuitively acceptable clustering

```
n=100; p=2; K=7
X=randn(n,p)
i=1
for d in ["complete","single","centroid","average"]:
    cluster=hc(X,dd=d)
    plt.subplot(2,2,i)
    i=i+1
    grp=cluster[K-1]
    for k in range(K):
        x=X[grp[k],0]
        y=X[grp[k],1]
        plt.scatter(x,y,s=5)
    plt.text(-2,2.1,"{}".format(d),fontsize=12)
```

Hierarchical clustering does not require the number of clusters K to be decided in advance. Compared to K-means clustering, in this case, the samples in each cluster are not close (Fig. 10.4). This result appears to be due to the phenomenon that hierarchical clustering is initially locally connected, and relatively distant samples were connected earlier after going to the higher layers. In addition, complete linkage is often used as a measure of the distance between clusters, depending on the application. The results in this case are closer to K-means clustering and are intuitively reasonable.

The result of hierarchical clustering is represented by a dendrogram (tree diagram) (Fig. 10.6). The shorter the distance of the cluster is, the earlier the

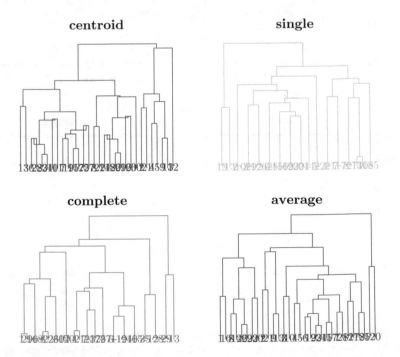

Fig. 10.6 Hierarchical clustering was executed for the artificial data with $N = 30$ and $p = 3$ using centroid, single, complete, and average linkages. The resulting dendrogram is displayed

clusters are connected, which means that the distance tends to increase as the process proceeds.

A tree is constructed so that the distance of the connected clusters is represented by the height of the branch at which they are connected. The higher we go, the fewer the branches. Then, for any $2 \le k \le N$, there is a height with the number of branches being k. If we cut the dendrogram horizontally at one of the heights, we obtain the clustering. The samples under the k branches consist of the k clusters.

However, when constructing a dendrogram, the samples in connected clusters should not cross, i.e., the branches should not intersect. It is necessary to arrange the samples at the lowest level so that this rule is not violated.

For single linkage, although the distance between the clusters is small in the early stages, the distance often rapidly increases after the connections become higher. For centroid linkage, which is often used in biology-related fields, inversion occurs such that two clusters are connected later than more distant pairs (Fig. 10.7).

Example 84 Suppose that we apply centroid linkage to the samples $(0, 0)$, $(5, 8)$, $(9, 0)$ with $N = 3$ and $p = 2$ (Fig. 10.8). Then, after the first connection, we obtain

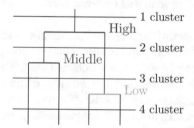

Fig. 10.7 The height, the number of clusters, and the distance between clusters for hierarchical clustering. The cluster distance indicated by red is larger than those indicated by blue and green. In addition, there are $k = 1, 2, 3, \ldots$ branches at each height below the root of the tree, but the cluster that consists of samples under a branch constitutes a cluster

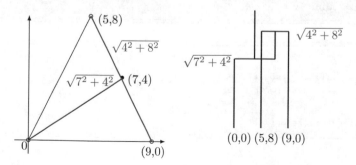

Fig. 10.8 Inversion for centroid linkage. The center of the blue cluster that connects $(5, 8)$ and $(9, 0)$ (the shortest among the three edge lengths $\sqrt{89}, 9, \sqrt{80}$) is $(7, 4)$. However, the distance of the red line between $(0, 0)$ and $(7, 4)$, which are connected later, is smaller. The dendrogram on the right displays the distance of connected clusters by the height of the branch: the red branch is lower than the branch of the already connected clusters, and they cross in the dendrogram

two clusters $\{(0, 0)\}$ and $\{(5, 8), (9, 0)\}$ with the cluster distance $\sqrt{4^2 + 8^2} = \sqrt{80}$. The distance between the centers $(0, 0)$ and $(7, 4)$ is $\sqrt{7^2 + 4^2} = \sqrt{65}$ in the next connection, which is smaller than the distance for the previous connection. This is an example of inversion, in which the branches cross in a dendrogram, as in Fig. 10.6 (lower left: centroid linkage).

However, no inversion occurs for complete, single, and average linkages: the later the connection is, the larger the distance between the connected clusters. In fact, from the definitions of the distances of clusters A and B for complete linkage in Table 10.1, because the two clusters to be connected have the largest cluster distance among the current clusters, no future cluster distance will be lower than the current distance. For single linkage, if clusters A and B connect and the pair A, B becomes a new cluster, the distance between the pair A, B and the other cluster C cannot be lowered. In fact, if such a phenomenon occurs, either A, C or B, C should have been connected prior to A, B, which contradicts the rule of single linkage to choose the clusters that minimize the minimum sample distance. Additionally, for average linkage, if the average distance between the pair A, B and the other cluster C is smaller than that between A and B, either the distance between A and C or that between B and C is smaller than that between A and B, which contradicts the rule of average linkage.

We generate a dendrogram based on the considerations in this section and list the code in the Appendix of this chapter because it may be complicated to follow in the main text. The dendrogram in Fig. 10.6 is obtained via the program.

Alternatively, we can use the function `scipy.cluster.hierarchy` prepared for the Python language. For $X \in \mathbb{R}^{N \times p}$, via `dist` function, we obtain the matrix in which the distances among N samples with p variables are in the lower left positions. We specify the matrix and the option `method = "complete"`, `"single"`, `"centroid"`, `"average"`.

Example 85

```
from scipy.cluster.hierarchy import linkage,dendrogram
```

```
X=randn(20,2)
i=1
for d in ["single","average","complete","weighted"]:
    res_hc=linkage(X,method=d)
    plt.subplot(2,2,i)
    i+=1
    dendrogram(res_hc)
```

10.3 Principal Component Analysis

The principal component analysis (PCA) is the procedure used to obtain p vectors
$\phi_1 \in \mathbb{R}^p$ with $\|\phi_1\| = 1$ that maximizes $\|X\phi_1\|$,
$\phi_2 \in \mathbb{R}^p$ with $\|\phi_2\| = 1$ that is orthogonal to ϕ_1 and maximizes $\|X\phi_2\|, \cdots,$
$\vdots$
$\phi_p \in \mathbb{R}^p$ with $\|\phi_p\| = 1$ that is orthogonal to $\phi_1, \cdots, \phi_{p-1}$ and maximizes $\|X\phi_p\|$.

from a data matrix $X \in \mathbb{R}^{N \times p}$ ($p \leq N$). Before performing the PCA, we often centralize the matrix X, i.e., we subtract the arithmetic mean $\overline{x}_j$ of column j from each element $x_{i,j}$, $i = 1, \cdots, N$, $j = 1, \cdots, p$.

The purpose of PCA is to summarize the matrix X as $\phi_1, \cdots, \phi_m$ ($1 \leq m \leq p$): the smaller the m is, the more compressed the information in X. We note that there exists μ_1 such that

$$X^T X \phi_1 = \mu_1 \phi_1 . \tag{10.3}$$

In fact, ϕ_1 maximizes $\|X\phi_1\|$, and if we differentiate

$$L := \|X\phi_1\|^2 - \gamma(\|\phi_1\|^2 - 1)$$

with a Lagrange constant γ and set it equal to zero, we find $\|\phi_1\|^2 = 1$ and $X^T X \phi_1 = \gamma \phi_1$. Although more than one μ_1 may satisfy (10.3), from

$$\|X\phi_1\|^2 = \phi_1^T X^T X \phi_1 = \mu_1 \|\phi_1\|^2 = \mu_1,$$

we need to choose the largest μ_1.

Additionally, for ϕ_2,

$$X^T X \phi_2 = \mu_2 \phi_2$$

is required for some μ_2. Hence, ϕ_2 is the eigenvector of $X^T X$ as well. If we note that $\mu_1 \geq \mu_2$ and that ϕ_1 and ϕ_2 are orthogonal, the possibility is either $\mu_1 = \mu_2$ (ϕ_1 and ϕ_2 are in the same eigenspace) or μ_2 is the largest but μ_1. Moreover, they are nonnegative because they are eigenvalues of a nonnegative definite matrix.

In the actual PCA formulation, we define $\Sigma := \frac{1}{N} X^T X$, replace $\mu_1/N, \ldots, \mu_m/N$ by $\lambda_1, \ldots, \lambda_N$, and write (10.3) as

$$\Sigma \phi_1 = \lambda_1 \phi_1,$$

where Σ is the sample-based covariance matrix and $\lambda_1 \geq \cdots \geq \lambda_p \geq 0$ are the eigenvalues.

We choose the m principal components $\phi_1, \ldots, \phi_m$ with the largest variances $\lambda_1 \geq \cdots \geq \lambda_m \geq 0$.

We say that $\dfrac{\lambda_k}{\sum_{i=1}^{p} \lambda_i}$ and $\dfrac{\sum_{i=1}^{k} \lambda_i}{\sum_{i=1}^{p} \lambda_i}$ are the proportion of the k-th principal component and the accumulated proportion up to the k-th principal components, respectively.

If the units of the p columns in X are different, we often need to scale X such that the result of PCA does not depend on the units. For example, if each column expresses the test score of math, English, science, etc., they may not have to be scaled, even if the variances are different. However, if each column expresses height, weight, age, etc., they may have to be scaled because if we replace centimeters, kilograms, and years by inches, pounds, and months, then the PCA produces significantly different results.

If the dimension of an eigenspace is not one, the only constraints are that the basis is orthogonal and the length is one. However, if the dimension is one, the eigenvector is either a vector or its oppositely directed vector.

When the matrix X is randomly generated, it is unlikely that more than one eigenvalue coincide, so we assume

$$\lambda_1 > \cdots > \lambda_m .$$

Since Σ is symmetric, we have $\lambda_i \neq \lambda_j \implies \phi_i^T \phi_j = 0$. Moreover, from

$$\lambda_j \phi_i^T \phi_j = \phi_i^T \Sigma \phi_j = \phi_j^T \Sigma \phi_i = \lambda_i \phi_i^T \phi_j ,$$

we have $(\lambda_i - \lambda_j)\phi_i^T \phi_j = 0$, which means $\phi_i^T \phi_j = 0$. We note that if we find the m largest eigenvalues and their eigenvectors, we do not have to check whether those eigenvectors are orthogonal.

Using the Python language function $\texttt{np.linalg.eig}$, given the matrix $X \in \mathbb{R}^{N \times p}$ as an input, we can construct the function $\texttt{pca}$ that outputs the vectors with the elements $\lambda_1, \ldots, \lambda_p$ and the matrix with the columns $\phi_1, \ldots, \phi_p$.

```
def pca(X):
    n,p=X.shape
    center=np.average(X,0)
    X=X-center # Centralization by column
    Sigma=X.T@X/n
    lam,phi=np.linalg.eig(Sigma)   # eigen values , eigen vectors
    index=np.argsort(-lam)   # Sort by descending order
    lam=lam[index]
    phi=phi[:,index]
    return {'lam':lam,'vectors':phi,'centers':center}
```

Even if we do not use the above function, the function $\texttt{PCA}$ in the $\texttt{sklearn.decomposition}$ is available for the Python language.

Example 86 We do not distinguish the two directions (a vector or its (-1) multiplication) of each of the principal component vectors, although they depend on the software.

```
X=randn(100,5)
res=pca(X)
res['lam']
```

```
array([110.53492367, 103.30322442,  94.67566385,  78.62762373,
        71.98586376])
```

```
array([0.24075006, 0.22499909, 0.20620787, 0.17125452, 0.15678846])
```

```
res['vectors']
```

```
array([[ 0.1904871 ,  0.86655739,  0.23631724,  0.34643019, -0.19218023],
       [ 0.65407668,  0.09134685, -0.59040129, -0.35265467, -0.30149701],
       [-0.13324667, -0.20604928, -0.50496326,  0.78034922, -0.27542008],
       [-0.5430764 ,  0.44470055, -0.57750325, -0.22518257,  0.35084505],
       [ 0.47245286, -0.02278504, -0.08415809,  0.30978817,  0.82049853]])
```

```
res['centers']
```

```
from sklearn.decomposition import PCA
```

```
pca=PCA()
pca.fit(X) # execute
```

```
PCA(copy=True, iterated_power='auto', n_components=None, random_state=None,
    svd_solver='auto', tol=0.0, whiten=False)
```

```
score=pca.fit_transform(X) # PC score ((rows: n, columns: PC score)
score[0:5,]
```

```
array([[-0.20579722,  0.63537368,  1.20127757, -0.17642322,  0.08331289],
       [ 1.81876319,  0.7014673 , -0.76877222,  0.94195901,  1.32429876],
       [-1.64856653,  1.27063092, -1.36066169, -0.0763228 , -0.81823956],
       [-1.01126137, -0.21633468,  1.21589032, -0.54061369,  0.14468562],
       [-0.71078308,  0.74867317,  0.81140784, -0.45036742, -0.27535244]])
```

```
array([[ 0.1904871 ,  0.65407668, -0.13324667, -0.5430764 ,  0.47245286],
       [ 0.86655739,  0.09134685, -0.20604928,  0.44470055, -0.02278504],
       [ 0.23631724, -0.59040129, -0.50496326, -0.57750325, -0.08415809],
       [-0.34643019,  0.35265467, -0.78034922,  0.22518257, -0.30978817],
       [ 0.19218023,  0.30149701,  0.27542008, -0.35084505, -0.82049853]])
```

```
pca.mean_     # it is same as above "centers"
```

```
array([-0.03670141,  0.03260174,  0.13786866,  0.00316844, -0.12808206])
```

We compute the proportions and the accumulated proportion in Fig. 10.9 (left).

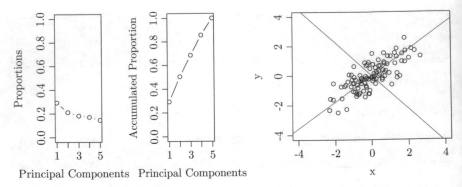

Fig. 10.9 Proportions and their accumulated proportion (left) and the mutually orthogonal first and second principal components (right)

```
plt.plot(np.arange(1,6),evr)
plt.scatter(np.arange(1,6),evr)
plt.xticks(np.arange(1,6))
plt.ylim(0,1)
plt.xlabel("principal_component")
plt.ylabel("contribution_rate")
```

```
Text(0, 0.5, 'contribution rate')
```

```
plt.plot(np.arange(1,6),np.cumsum(evr))
plt.scatter(np.arange(1,6),np.cumsum(evr))
plt.xticks(np.arange(1,6))
plt.ylim(0,1)
plt.xlabel("principal_component
")
plt.ylabel("contribution_rate")
```

```
Text(0, 0.5, 'contribution rate')
```

Example 87 Given N observations $(x_1, y_1), \ldots, (x_N, y_N)$, we wish to find the mutually orthogonal principal components ϕ_1 and ϕ_2.

```
n=100; a=0.7; b=np.sqrt(1-a**2)
u=randn(n); v=randn(n)
x=u; y=u*a+v*b
plt.scatter(x,y); plt.xlim(-4,4); plt.ylim(-4,4)
```

```
(-4, 4)
```

```
D=np.concatenate((x.reshape(-1,1),y.reshape(-1,1)),1)
pca.fit(D)
```

```
PCA(copy=True, iterated_power='auto', n_components=None, random_state=None,
    svd_solver='auto', tol=0.0, whiten=False)
```

```
T=pca.components_
T[0,1]/T[0,0]*T[1,1]/T[1,0]     # PC vectors   are orthogonal
```

```
-1.0
```

```
def f_1(x):
    y=T[0,1]/T[0,0]*x
    return y
```

```
def f_2(x):
    y=T[1,1]/T[1,0]*x
    return y
```

```
x_seq=np.arange(-4,4,0.5)
plt.scatter(x,y,c="black")
plt.xlim(-4,4)
plt.ylim(-4,4)
plt.plot(x_seq,f_1(x_seq))
plt.plot(x_seq,f_2(x_seq))
plt.gca().set_aspect("equal",adjustable="box")
```

```
(-4.375, 3.875, -4.4445676982833735, 5.018060304513487)
```

Note that the product of the two lines is -1 (Fig. 10.9, right).

Using the obtained $\phi_1, \ldots, \phi_m$ and their projections $z_1 = X\phi_1, \ldots, z_m = X\phi_m$, we can see the N data projected on the m-dimensional space, and the function `biplot` is available for this purpose.

Example 88 A dataset containing the numbers of arrests for four crimes in all fifty states is available. We scale the data so that the four variances are equal, execute PCA, and plot the first and second principal components. The function like `biplot` in R is not available in the Python language for this purpose, so we make the function. Because we project the data onto two dimensions, our analysis considers the first two components ($m = 2$). If we multiply the first and second principal components by -1, we obtain principal component vectors that have the same variance but the opposite direction and projection values (Fig. 10.10)

```
import pandas as pd
```

```
USA=pd.read_csv('USArrests.csv',header=0,index_col=0)
X=(USA-np.average(USA,0))/np.std(USA,0)
index=USA.index
col=USA.columns
pca=PCA(n_components=2)
pca.fit(X)
score=pca.fit_transform(X)
vector=pca.components_
vector
```

Fig. 10.10 Using the function `biplot`, we project the data on the crimes in the fifty states on the first and second components

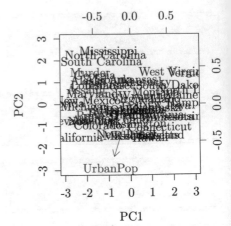

```
array([[ 0.53589947,  0.58318363,  0.27819087,  0.54343209],
       [ 0.41818087,  0.1879856 , -0.87280619, -0.16731864]])
```

```
vector.shape[1]
```

```
4
```

```
evr=pca.explained_variance_ratio_
evr
```

```
array([0.62006039, 0.24744129])
```

```
plt.figure(figsize=(7,7))
for i in range(score.shape[0]):
    plt.scatter(score[i,0],score[i,1],s=5)
    plt.annotate(index[i],xy=(score[i,0],score[i,1]))
for j in range(vector.shape[1]):
    plt.arrow(0,0,vector[0,j]*2,vector[1,j]*2,color="red")  #  2 is the
        length of the line, you can choose arbitrary.
    plt.text(vector[0,j]*2,vector[1,j]*2,col[j],color="red")
```

The principal component analysis is used to reduce the dimensionality of multivariate data. The clustering learned in the previous section cannot be displayed unless the data are two-dimensional. A possible method is to display samples after reducing the space to two dimensions via principal component analysis.

Example 89 We display the output of the K-means clustering of the Boston data as a two-dimensional principal component (Fig. 10.11). Since the data are projected in two dimensions, when viewed as a two-dimensional graph, it does not appear that close samples consist of a cluster.

Fig. 10.11 We projected the K-means clustering results of the Boston data on the first and second components

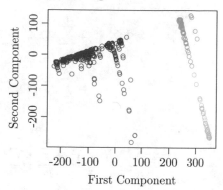

```
from sklearn.datasets import load_boston
from sklearn.cluster import KMeans
```

```
Boston=load_boston()
Z=np.concatenate((Boston.data,Boston.target.reshape(-1,1)),1)
K_means=KMeans(n_clusters=5)
K_means.fit(Z)
y=K_means.fit_predict(Z)   # predict which cluster
pca.fit(Z)
W=pca.fit_transform(Z)[:,[0,1]] # The first and second principal components
    for each n
plt.scatter(W[:,0],W[:,1],c=y)
plt.xlabel("_first_PC_component")
plt.ylabel("_second_PC_component")
plt.title("_clustering_with_Boston_data")
```

```
Text(0.5, 1.0, 'clustering with Boston data')
```

There is another equivalent definition of PCA. Suppose we centralize the matrix $X \in \mathbb{R}^{N \times p}$ and let x_i be the i-th row vector of $X \in \mathbb{R}^{N \times p}$ and $\Phi \in \mathbb{R}^{p \times m}$ be vectors such that the columns $\phi_1, \ldots, \phi_m$ have unit length and are mutually orthogonal. Then, we obtain the projections $z_1 = x_1 \Phi, \ldots, z_N = x_N \Phi \in \mathbb{R}^m$ (row vectors) from $x_1, \ldots, x_N$ to $\phi_1, \ldots, \phi_m$. We evaluate how close the recovered vectors are to the original $x_1, \ldots, x_N$ by

$$L := \sum_{i=1}^{N} \|x_i - x_i \Phi \Phi^T\|^2, \tag{10.4}$$

which is obtained by multiplying $z_1, \ldots, z_N$ by Φ^T from the right. If $m = p$, the value of (10.4) is zero. We may regard PCA as the problem of finding $\phi_1, \ldots, \phi_m$

that minimize (10.4). In fact, we have the following two equations:

$$\sum_{i=1}^{N} \|x_i - x_i \Phi\Phi^T\|^2 = \sum_{i=1}^{N} \{\|x_i\|^2 - 2x_i(x_i\Phi\Phi^T)^T + (x_i\Phi\Phi^T)(x_i\Phi\Phi^T)^T\}$$

$$= \sum_{i=1}^{N} \{\|x_i\|^2 - x_i\Phi\Phi^T x_i^T\} = \sum_{i=1}^{N} \|x_i\|^2 - \sum_{i=1}^{N} \|x_i\Phi\|^2$$

$$\sum_{i=1}^{N} \|x_i\Phi\|^2 = \sum_{i=1}^{N}\sum_{j=1}^{m}(x_i\phi_j)^2 = \sum_{j=1}^{m}\left\{\sum_{i=1}^{N}(x_i\phi_j)^2\right\} = \sum_{j=1}^{m}\|\begin{bmatrix} x_1\phi_j \\ \vdots \\ x_N\phi_j \end{bmatrix}\|^2 = \sum_{j=1}^{m}\|X\phi_j\|^2.$$

In other words, if $\lambda_1, \ldots, \lambda_m$ are the largest m eigenvalues of $\Sigma = X^T X/N$, the value $\sum_{i=1}^{N} \|x_i - x_i\Phi\Phi^T\|^2$ takes the minimum value $\sum_{i=1}^{N} \|x_i\|^2 - \sum_{j=1}^{m} \lambda_j$ by the m largest eigenvalues whose eigenvectors are $\phi_1, \ldots, \phi_m$.

In addition to PCA and linear regression, we may use principal component regression: find the matrix $Z = X\Phi \in \mathbb{R}^{N \times m}$ that consists of the m principal components via PCA, find $\theta \in \mathbb{R}^m$ that minimizes $\|y - Z\theta\|^2$, and display via $\hat{\theta}$ the relation between the response and m components (a replacement of the p covariates). Principal component regression regresses y on the columns of Z instead of those of X.

For $m = p$, $\Phi\hat{\theta}$ and $\hat{\beta} = (X^T X)^{-1}X^T y$ coincide. In fact, since $\min_\beta \|y - X\beta\|^2 \leq \min_\theta \|y - X\Phi\theta\|^2 = \min_\theta \|y - Z\theta\|^2$, the matrix Φ is nonsingular when $p = m$. Thus, for arbitrary $\beta \in \mathbb{R}^p$, there exists a θ such that $\beta = \Phi\theta$. For example, we may construct the following program:

```
def pca_regression(X,y,m):
    pca=PCA(n_components=m)
    pca.fit(X)
    Z=pca.fit_transform(X)   # rows:n , columns:PC score
    phi=pca.components_  #  rows : PC , columns : variables
    theta=np.linalg.inv(Z.T@Z)@Z.T@y
    beta=phi.T@theta
    return {'theta':theta,'beta':beta}
```

Example 90 We execute the function pca_regression:

```
n=100; p=5
X=randn(n,p)
X=X-np.average(X,0)
y=X[:,0]+X[:,1]+X[:,2]+X[:,3]+X[:,4]+randn(n)
y=y-np.mean(y)
pca_regression(X,y,3)
```

```
{'beta': array([1.33574835, 0.45612768, 0.6710805 , 0.28063559, 0.97748932]),
 'theta': array([ 0.41755766,  0.19389454, -1.80690824])}
```

```
pca_regression(X,y,5)['beta']
```

```
array([0.86513279, 1.01698307, 0.7496746 , 0.91010065, 1.12420093])
```

```
np.linalg.inv(X.T@X)@X.T@y
```

```
array([0.86513279, 1.01698307, 0.7496746 , 0.91010065, 1.12420093])
```

Appendix: Program

A program generates the dendrogram of hierarchical clustering. After obtaining the cluster object via the function hc, we compare the distances between consecutive clusters using the ordered sample y. Specifically, we express the positions of the branches by $z[k, 1]$, $z[k, 2]$, $z[k, 3]$, $z[k, 4]$, and $z[k, 5]$.

```
import matplotlib.pyplot as plt
import matplotlib.collections as mc
import matplotlib.cm as cm
```

```
def unlist(x):
    y=[]
    for z in x:
        y.extend(z)
    return(y)
```

```
def hc_dendroidgram(cluster,dd="complete",col="black"):
    y=unlist(cluster[0])
    n=len(y)
    z=np.zeros([n,5])
    index=[[y[i]] for i in range(n)]
    height=np.zeros(n)
    for k in range(n-1,0,-1):
        dist_min=np.inf
        for i in range(k):
            i_0=index[i]; j_0=index[i+1]
            if dd=="complete":
                d=dist_complete(X[i_0,],X[j_0,])
            elif dd=="single":
                d=dist_single(X[i_0,],X[j_0,])
            elif dd=="centroid":
                d=dist_centroid(X[i_0,],X[j_0,])
            elif dd=="average":
                d=dist_average(X[i_0,],X[j_0,])
            if d<dist_min:
                dist_min=d
                i_1=i    # list of index which would be combined
```

```
            j_1=j      # list of index which would join
    # below, calculate the position of the line segments
    i=0
    for h in range(i_1):
        i=i+len(index[h])
    z[k,0]=i+len(index[i_1])/2
    z[k,1]=i+len(index[i_1])+len(index[j_1])/2
    z[k,2]=height[i_1]
    z[k,3]=height[j_1]
    z[k,4]=dist_min
    index[i_1].extend(index[j_1])
    if j_1<k:                  #  put the added index forward
        for h in range(j_1,k):
            index[h]=index[h+1]
            height[h]=height[h+1]
    height[i_1]=dist_min
    height[k]=0
    # Loop ends here.
lines=[[(z[k,0],z[k,4]),(z[k,0],z[k,2])] for k in range(1,n)] # Vertical
    line segment (left)
lines2=[[(z[k,0],z[k,4]),(z[k,1],z[k,4])] for k in range(1,n)] #
    Horizontal line segment (center)
lines3=[[(z[k,1],z[k,4]),(z[k,1],z[k,3])] for k in range(1,n)] #
    Vertical line segment (right)
lines.extend(lines2)
lines.extend(lines3)
lc=mc.LineCollection(lines,colors=col,linewidths=1)
fig=plt.figure(figsize=(4,4))
ax=fig.add_subplot()
ax.add_collection(lc)
ax.autoscale()
plt.show()
fig=plt.figure(figsize=(4,4))
```

```
n=100; p=2; K=7
X=randn(n,p)
cluster=hc(X,dd="complete")
hc_dendroidgram(cluster,col="red")
```

Exercises 88–100

88. The following procedure divides N samples with p variables into K disjoint sets, given K (K-means clustering). We repeat the following two steps after randomly assigning one of $1, \ldots, K$ to each sample:

(a) Compute the centers of clusters $k = 1, \ldots, K$.
(b) To each of the N samples, assign the nearest center among the K clusters.

Fill in the blanks and execute the procedure.

```
def k_means(X,K,iteration=20):
    n,p=X.shape
    center=np.zeros((K,p))
    y=np.random.choice(K,n,replace=True)
    scores=[]
```

```
      for h in range(iteration):
          for k in range(K):
              if np.sum(y==k)==0:
                  center[k,0]=np.inf
              else:
                  for j in range(p):
                      center[k,j]=# blank(1) #
          S_total=0
          for i in range(n):
              S_min=np.inf
              for k in range(K):
                  S=np.sum((X[i,]-center[k,])**2)
                  if S<S_min:
                      S_min=S
                      # blank(2) #
              S_total+=S_min
          scores.append(S_total)
      return {'clusters':y,'scores':scores}
```

```
n=1000; K=5; p=2
X=randn(n,p)   # data generation
y=k_means(X,5)['clusters'] # getting cluster for each sample
plt.scatter(X[:,0],X[:,1],c=y)
plt.xlabel("first_component")
plt.ylabel("second_component")
```

89. The clusters that K-means clustering generates depend on the randomly chosen initial values. Repeat ten times to find the sequence of values immediately after the 2-step update. Display each transition as a line graph on the same graph.

90. K-means clustering minimizes

$$S := \sum_{k=1}^{K} \frac{1}{|C_k|} \sum_{i \in C_k} \sum_{i' \in C_k} \sum_{j=1}^{p} (x_{i,j} - x_{i',j})^2$$

w.r.t. $C_1, \ldots, C_K$ from data $X = (x_{i,j})$.

(a) Show the following equation:

$$\frac{1}{|C_k|} \sum_{i \in C_k} \sum_{i' \in C_k} \sum_{j=1}^{p} (x_{i,j} - x_{i',j})^2 = 2 \sum_{i \in C_k} \sum_{j=1}^{p} (x_{i,j} - \bar{x}_{k,j})^2.$$

(b) Show that the score S is monotonously decreasing each time the two steps are executed in Problem 88.

(c) Let $N = 3$, $p = 1$, and $K = 2$, and assume that the samples are in $0, 6, 10$. We consider two cases: one and two are assigned to $0, 6$ and 10, respectively, and one and two are assigned to 0 and $6, 10$, respectively. What values do they converge to if the initial state is each of the two cases? What score do they finally obtain?

91. Write Python codes for the functions dist_complete, dist_single, dist_centroid, and dist_average to find the maximum distance between the rows in x, y, the minimum distance between the rows in x, y, the distance between the centers of x, y, and the average distance between the rows in x, y, given matrices x and y composed of multiple rows extracted from $X \in \mathbb{R}^{N \times p}$.

92. The following procedure executes hierarchical clustering w.r.t. data $x_1, \ldots, x_N \in \mathbb{R}^p$. Initially, each cluster contains exactly one sample. We merge the clusters to obtain a clustering with any number K of clusters. Fill in the blanks and execute the procedure.

```
import copy
def hc(X,dd="complete"):
    n=X.shape[0]
    index=[[i] for i in range(n)]
    cluster=[[] for i in range(n-1)]
    for k in range(n,1,-1):
        # index_2=[]
        dist_min=np.inf
        for i in range(k-1):
            for j in range(i+1,k):
                i_0=index[i]; j_0=index[j]
                if dd=="complete":
                    d=dist_complete(X[i_0,],X[j_0,])
                elif dd=="single":
                    d=dist_single(X[i_0,],X[j_0,])
                elif dd=="centroid":
                    d=dist_centroid(X[i_0,],X[j_0,])
                elif dd=="average":
                    d=dist_average(X[i_0,],X[j_0,])
                if d<dist_min:
                    # blank(1) #
                    i_1=i    # list of index which would be combined
                    j_1=j    # list of index which would join
        index[i_1].extend(index[j_1])   # add
        if j_1<k:                        # put the added index forward
            for h in range(j_1+1,k,1):
                index[h-1]=# blank(2) #
        index2=copy.deepcopy(index[0:(k-1)])   # If you use "index"
            without deepcopy , "index" will be rewritten each time.
        cluster[k-2].extend(index2)
    return cluster  # The results from below show that one by one, the
        congruence occurs.
```

93. In hierarchical clustering, if we use centroid linkage, which connects the clusters with the smallest value of dist_centroid, inversion may occur, i.e., clusters with a smaller distance can be connected later. Explain the phenomenon for the case $(0, 0), (5, 8), (9, 0)$ with $N = 3$ and $p = 2$.

94. Let $\Sigma = X^T X / N$ for $X \in \mathbb{R}^{N \times p}$, and let λ_i be the i-th largest eigenvalue in Σ.

(a) Show that the ϕ that maximizes $\|X\phi\|^2$ among $\phi \in \mathbb{R}^N$ with $\|\phi\| = 1$ satisfies $\Sigma\phi = \lambda_1\phi$.

(b) Show $\phi_1, \ldots, \phi_m$ such that $\Sigma\phi_1 = \lambda_1\phi_1, \ldots$, and $\Sigma\phi_m = \lambda_m\phi_m$ are orthogonal when $\lambda_1 > \cdots > \lambda_m$.

95. Using the `np.linalg.eig` function in the Python language, write a Python program `pca` that outputs the average of the p columns, the eigenvalues $\lambda_1, \ldots, \lambda_p$, and the matrix that consists of $\phi_1, \ldots, \phi_p$, given input $X \in \mathbb{R}^{N \times p}$. Moreover, execute the following to show that the results obtained via PCA in `sklearn.decomposition` coincide:

```
X=randn(100,5)
res=pca(X)
res['lam']
res['vectors']
res['centers']
```

```
from sklearn.decomposition import PCA
```

```
pca=PCA()
pca.fit(X) # execute
score=pca.fit_transform(X) # PC score ((rows: n, columns: PC score)
score[0:5,]
pca.mean_    # it is same as above "centers"
```

96. The following procedure produces the first and second principal component vectors ϕ_1 and ϕ_2 from N samples $(x_1, y_1), \ldots, (x_N, y_N)$. Fill in the blanks and execute it.

```
n=100; a=0.7; b=np.sqrt(1-a**2)
u=randn(n); v=randn(n)
x=u; y=u*a+v*b
plt.scatter(x,y); plt.xlim(-4,4); plt.ylim(-4,4)
D=np.concatenate(((x.reshape(-1,1),y.reshape(-1,1)),1)
pca.fit(D)
T=pca.components_
T[1,0]/T[0,0]*T[1,1]/T[0,1]    # PC vectors are orthogonal
```

```
-1.0
```

```
def f_1(x):
    y=# blank(1) #
    return y
```

```
def f_2(x):
    y=T[1,1]/T[1,0]*x
    return y
```

```
x_seq=np.arange(-4,4,0.5)
plt.scatter(x,y,c="black")
plt.xlim(-4,4)
plt.ylim(-4,4)
plt.plot(x_seq,f_1(x_seq))
plt.plot(x_seq,# blank(2) #)
plt.gca().set_aspect("equal",adjustable="box")
```

Moreover, show that the product of the slopes is -1.

97. There is another equivalent definition of PCA. Suppose that we have centralized the matrix $X \in \mathbb{R}^{N \times p}$, and let x_i be the i-th row vector of $X \in \mathbb{R}^{N \times p}$ and $\Phi \in \mathbb{R}^{p \times m}$ be the matrix that consists of the mutually orthogonal vectors $\phi_1, \ldots, \phi_m$ of unit length. Then, we can obtain the projection $z_1 = x_1\Phi, \ldots, z_N = x_N\Phi \in \mathbb{R}^m$ of $x_1, \ldots, x_N$ on $\phi_1, \ldots, \phi_m$. We evaluate how the $x_1, \ldots, x_N$ are recovered by $L := \sum_{i=1}^{N} \|x_i - x_i\Phi\Phi^T\|^2$, which is obtained by multiplying $z_1, \ldots, z_N$ by Φ^T from the right. We can regard PCA as the problem of finding $\phi_1, \ldots, \phi_m$ that minimize the value. Show the two equations:

$$\sum_{i=1}^{N} \|x_i - x_i\Phi\Phi^T\|^2 = \sum_{i=1}^{N} \|x_i\|^2 - \sum_{i=1}^{N} \|x_i\Phi\|^2$$

$$\sum_{i=1}^{N} \|x_i\Phi\|^2 = \sum_{j=1}^{m} \|X\phi_j\|^2.$$

98. We prepare a dataset containing the numbers of arrests for four crimes in all fifty states.

```
import pandas as pd
```

```
USA=pd.read_csv('USArrests.csv',header=0,index_col=0)
X=(USA-np.average(USA,0))/np.std(USA,0)
index=USA.index
col=USA.columns
pca=PCA(n_components=2)
pca.fit(X)
score=pca.fit_transform(X)
vector=pca.components_
vector
vector.shape[1]
evr=pca.explained_variance_ratio_
evr
```

```
plt.figure(figsize=(7,7))
for i in range(score.shape[0]):
    plt.scatter(# blank(1) #,# blank(2) #,s=5)
    plt.annotate(# blank(3) #,xy=(score[i,0],score[i,1]))
for j in range(vector.shape[1]):
    plt.arrow(0,0,vector[0,j]*2,vector[1,j]*2,color="red")    #  2 is
        the length of the line, you can choose arbitrary.
    plt.text(vector[0,j]*2,vector[1,j]*2,col[j],color="red")
```

Fill in the blanks and execute the following code:

99. The proportions and accumulated proportion are defined by $\dfrac{\lambda_k}{\sum_{j=1}^{p} \lambda_j}$ and

$\dfrac{\sum_{k=1}^{m} \lambda_k}{\sum_{j=1}^{p} \lambda_j}$ for each $1 \leq m \leq p$. Fill in the blanks and draw the graph.

```
res['lam']/np.sum(res['lam'])   #  Contributions of each principal
    component
```

```
evr=pca.explained_variance_ratio_   #
evr
```

```
plt.plot(np.arange(1,6),evr)
plt.scatter(np.arange(1,6),evr)
plt.xticks(np.arange(1,6))
plt.ylim(0,1)
plt.xlabel("principal component ")
plt.ylabel("contribution rate")
```

```
plt.plot(np.arange(1,6),np.cumsum(evr))
plt.scatter(np.arange(1,6),np.cumsum(evr))
plt.xticks(np.arange(1,6))
plt.ylim(0,1)
plt.xlabel("principal component ")
plt.ylabel("contribution rate")
```

100. In addition to PCA and linear regression, we may use principal component regression: find the matrix $Z = X\Phi \in \mathbb{R}^{N \times m}$ that consists of the m principal components obtained via PCA, find $\theta \in \mathbb{R}^m$ that minimizes $\|y - Z\theta\|^2$, and display via $\hat{\theta}$ the relation between the response and m components (a replacement of the p covariates). Principal component regression regresses y on the columns of Z instead of those of X.

Show that $\Phi\hat{\theta}$ and $\beta = (X^T X)^{-1} X^T y$ coincide for $m = p$. Moreover, fill in the blanks and execute it.

```
def pca_regression(X,y,m) :
    pca=PCA(n_components=m)
    pca.fit(X)
    Z=pca.fit_transform(X)   # rows:n, columns: PCscore
    phi=pca.components_   # rows:PC, columns: variables
    theta=# blank #
    beta=phi.T@theta
    return {'theta':theta,'beta':beta}
```

Hint: Because $\min_\beta \|y - X\beta\|^2 \leq \min_\theta \|y - X\Phi\theta\|^2 = \min_\theta \|y - Z\theta\|^2$, it is sufficient to show that there exists θ such that $\beta = \Phi\theta$ for an arbitrary $\beta \in \mathbb{R}^p$ when $p = m$.

Correction to: Unsupervised Learning

Correction to:
Chapter 10 in: J. Suzuki, *Statistical Learning with Math and*
Python, **https://doi.org/10.1007/978-981-15-7877-9_10**

In the original version of the book, the following corrections have been incorporated: The phrases "principle component" and "principle component analysis" have been changed to "principal component" and "principal component analysis", respectively, at all occurrences in Chap. 10. The book and the chapter have been updated with the changes.

The updated original version for this chapter can be found at
https://doi.org/10.1007/978-981-15-7877-9_10

Index

© The Author(s), under exclusive license to Springer Nature Singapore Pte Ltd. 2021
J. Suzuki, *Statistical Learning with Math and Python*,
https://doi.org/10.1007/978-981-15-7877-9

Printed in the United States
by Baker & Taylor Publisher Services